Human–Systems Integration

Human–Systems Integration

From Virtual to Tangible

Guy André Boy

CRC Press

Taylor & Francis Group

Boca Raton   London   New York

CRC Press is an imprint of the
Taylor & Francis Group, an **informa** business

CRC Press
Taylor & Francis Group
6000 Broken Sound Parkway NW, Suite 300
Boca Raton, FL 33487-2742

© 2020 by Taylor & Francis Group, LLC
CRC Press is an imprint of Taylor & Francis Group, an Informa business

No claim to original U.S. Government works

Printed on acid-free paper

International Standard Book Number-13: 978-0-367-36841-8 (Hardback)
International Standard Book Number-13: 978-0-367-35773-3 (Paperback)

Visit the Taylor & Francis Web site at
http://www.taylorandfrancis.com

and the CRC Press Web site at
http://www.crcpress.com

Contents

Preface

> To exist is to change, to change is to mature, to mature is to go on creating oneself endlessly.

Henri Bergson
Philosopher

This book is an attempt to better formalize a systemic approach to Human–Systems Integration (HSI). Good HSI is a matter of maturity… it takes time to mature. It takes time for a human being to become autonomous and then mature! HSI is a matter of human–machine teaming, where human–machine cooperation and coordination are crucial. We cannot think "engineering design" without considering people and organizations that go with it. We also cannot think "new technology," "new organizations," and "new jobs" without considering change management.

HSI is interdisciplinary. It involves participation of life, human, and social sciences (LHSS) together with STEM[1] disciplines. In addition, HSI should involve creativity to shift from STEM to STEAM, where the "A" is for Arts. Indeed, HSI should combine divergent and convergent thinking to open the field of design possibilities and close it at the same time considering constraints from physics and LHSS. HSI is still a nascent practice that requires more conceptualization regarding approaches, methods, and tools. Making an interdisciplinary program is always a very difficult challenge.

This book is a follow-up of previous contributions in **human-centered design (HCD)** (Boy, 2013a) and practice in the development of virtual prototypes that requires progressive operational tangibility (Boy, 2016). HCD is an interdisciplinary approach that mainly mixes cognitive engineering, advanced human–computer interaction (HCI), modeling and simulation, complexity management, life-critical systems, and organization design and management. Interdisciplinary practice is both artistic and rational,

---

[1] Science, technology, engineering, and mathematics.

in the sense that it requires creativity (i.e., divergent thinking) and structured reasoning (i.e., convergent thinking).

HSI is a very touchy domain. This is because it belongs to several disciplines, and at the same time, it is a new endeavor for everybody on this planet. Human factors and ergonomics (HFE) specialists say that HSI is their thing because they have already developed approaches and methods to handle it. For them, HSI is a continuity of HFE, full stop! However, starting in the 1980s, HCI community developed and created new approaches leading to interaction design and HCD. This community focused on computing and more specifically on office automation, telecommunication systems, such as smart phones, computer-supported cooperative work, and web-based systems. Industrial complex systems were not on HCI community radar, but HCD of everyday things was (Norman, 2013). More recently, the systems engineering (SE) community became aware that HSI was a crucial topic, since people and organizations are integrating parts of systems of systems. Virtual HCD (VHCD) has become a reality that helps supporting HFE, HCI, and SE at design time and all along the life cycle of a system. HSI should be understood as the result of the association of HFE, HCI, VHCD, and SE. Since VHCD is now used in many industrial settings, it deserves more attention in terms of tangibility (i.e., making sure that designed systems are humanly graspable in the physical sense as well as the figurative sense). In other words, systems should be humanly easy to manipulate and understand.

Since my return to France in late 2017 after many years in Florida, I managed to set up a joint chair program and a new lab at ESTIA[2] Institute of Technology (Basque Country) and CentraleSupélec (Paris Saclay University). Today, FlexTech Chair and CLE (Concept Lab ESTIA) have started their development and support of several HSI research and innovation projects. I would like to thank people who helped me putting together this program and lab. Among them are primarily Patxi Elissalde, Bernard Yannou, and Jean-Roch Guiresse, who were early contributors to the success of FlexTech. Thank you also to Jean-Patrick Gaviard, Olivier Giquel, Hélène Huard de la Marre, Cynthia Lamothe, and Alexandrine Urbain, who provided major support to the overall endeavor. Thanks to Dassault Systèmes for their support in the initial setup of CLE.

Many people contributed to make this book a reality through long discussions, controversies, and agreements. Many thanks to Audrey Abi Akle, David Atkinson, Rudolph Balciunas, Thierry Baron, Deborah Boehm-Davis, Marie-Catherine Beaudoux, Thierry Bellet, Michael Boardman, Sébastien Boulnois, Jeremy Boy, Stélian Camara Dit Pinto, Sharon Chinoy,

---

[2] École Supérieure des Technologies Industrielles Avancées, a French engineering school. Estia means "home" in Greek… this may be the reason why I subconsciously chose to be there!

Bernard Claverie, Nadine Couture, Brigitte Daniel, Françoise Darses, Ken Davidian, Bernardo Delicado, Julien Dezemery, Emmanuel Dufrasne, Francis Durso, Jean-Jacques Favier, Alain Garcia, Jean-Patrick Gaviard, Eapen George, Daniel Hauret, Daniel Krob, Justin Larouzé, Olivier Larre, Jérémy Legardeur, Larry Leifer, Ludovic Loine, Dimitri Masson, Kathleen Mosier, Jean-Michel Munoz, Mark Musen, Donald Norman, Cynthia Null, Philippe Palanque, Edwige Quillerou-Grivot, David Pappalardo, Jean Pinet, Jérôme Ranc, Liya Regel, Garry Roedler, Chloé Rolos, Alexander Rudolph, Anabela Simoes, François Thermy, Eric Villeneuve, Terry Winograd, and Avigdor Zonnenshain.

Guy André Boy
Biarritz, September 18, 2019

Author

The author of the book is FlexTech Chair Institute Professor of Human–Systems Integration (HSI) at CentraleSupélec and ESTIA Institute of Technology in collaboration with several industrial partners. This work is immersed in the emerging field of operations and organizations digitalization, addressed by many other research teams. The underlying vision is human-centered during the whole life cycle of systems being investigated. Our HSI approach is strongly embedded into the aerospace sector and, more generally, in the field of complex systems-of-systems automation, where the term "system" denotes either humans or machines, and most importantly association of both.

In parallel, HSI is currently developed within an INCOSE (International Council on Systems Engineering) working group, the HSI Working Group, chaired by the author of the book, and populated with scientists and practitioners from industry (e.g., Airbus, Boeing, Lockheed Martin, Shell, Northrop, Ford, AREVA), government agencies (e.g., NASA, JPL, MOD, DOD), and academic research (e.g., Georgia Tech, FIT, Cranfield, Stevens, CentraleSupélec). INCOSE offers the state of the art on SE for the past three decades to improve productivity and effectiveness of industrial processes and standards. INCOSE gathers more than 17,000 members, distributed among seventy chapters over more than thirty-five countries around the globe. The first INCOSE HSI conference was held in Biarritz, France in September 2019.

chapter zero

Introduction

> Photography is a kind of virtual reality, and it helps
> if you can create the illusion of being in an interest-
> ing world.
>
> **Steven Pinker**
> *Professor of Psychology at Harvard University*

In this book, I would like to use this "photography" quote from Steven
Pinker to "modeling and simulation" and target several audiences that
include industrial and systems engineering (SE), engineering design,
human factors and ergonomics, human and social sciences, and management of all kinds. It presents an integrative approach that combines
human-centered design (HCD)[1] and SE, leading toward Human–Systems
Integration (HSI) (Boy & Narkevicius, 2013). It is a follow-up of two books
that I previously published, *Orchestrating Human-Centered Design* in 2013
and *Tangible Interactive Systems* in 2016. It is also a vision that is shared by
INCOSE HSI Working Group, which is working on the refinement of such
a definition of HSI and has the goal to gather interdisciplinary contributions. In this book, HSI will be considered as both a process and a product,
at the intersection of several disciplines.

What did we learn from the Apollo 13's "successful failure"?

HSI is a matter of technology, organization, and people (TOP). I would
say people first! However, people should be in a well-organized setup
and have necessary and sufficient technology to handle a large variety of
situations. Apollo program is an excellent example of HSI for satisfying
a very difficult goal, that is, doing a round trip to the Moon. People who
have made Apollo program a reality worked for several years together to
build a rocket, a Moon orbiter and a Moon lander, and finally missions

---

[1] The term "HCD" is often discussed by many people who propose other terms such as
Human-Centered Engineering, Human–Systems Integration, and Ergonomic Design.
Confusions come from the fact that we do not have the same background to coin a term
for a concept. Nevertheless, HCD is commonly accepted and taught in several academic
and professional programs worldwide.

(from preparatory missions to effective missions where they landed on the Moon with Apollo 11 in July 1969 for the first time, to Apollo 17 in December 1972). Several simulations have been implemented prior to taking the risk to land on the Moon. It has been an incremental *tangibilization* of imagined technological, organizational, and human concepts based on physics and a solid teaming effort.

Of course, in this kind of endeavor, not everything works as expected. In April 1970, the Apollo 13 mission did not go well. Suddenly, a message was sent by the crew to mission control: "Houston, we've had a problem." This was the start of a cooperation between orbiter Odyssey, attached to lander Aquarius, altogether the Apollo 13 spacecraft, and Houston Mission Control when the crew figured out an alarm indicating that oxygen pressure fell and power disappeared. The malfunction was caused by an explosion and rupture of an oxygen tank.

The astronauts had moved to Aquarius (the lunar module) from Odyssey (the command module) to keep power for the emergency return to Earth. They had lithium hydroxide canisters to cleanse their spacecraft of carbon dioxide, but some of the backup square canisters were not compatible with the round openings in the lunar module. In the meantime, Houston engineers put together a plan using plastic bags, cardboard, and duct tape to retrofit the canisters. It took more than a day to build a mock-up and get instructions to the crew. This was "problem solving" by a team of experts, which made a repair procedure that they sent to the crew. The crew used the procedure, put together the system onboard, and it worked out fine!

It is interesting to see how a possible disaster could be transformed into a successful operation. Why? They were all very well prepared and trained to deal with risk. They succeeded to solve an unexpected problem that was very complex and almost not formalized. They used creativity, knowledge, simulation, rapid prototyping, and cooperation among various experts to create an ad hoc solution. Of course, some people can say that it was luck, but it was really proactive engagement of motivated and knowledgeable people coordinated through collaborative problem solving. It was also a goal-driven project with a reasonable budget!

As already said, TOP matter. Many years after, during the HCI-Aero 2010 conference, I had the chance to work with Gerald Griffin, a former NASA Flight Director (Apollo 14, 16, and 17), who told me the following: "Apollo program ground actors were mostly engineers and science graduates. Many had civilian and/or military aviation backgrounds. They were young. They always considered themselves as extensions of crew and spacecraft. They absorbed training like sponges. They had high respect for flight crews (work and other). Their greatest strengths were situational awareness and fear of screwing up!"

In this book, we will see how good HSI could make a difference in normal, abnormal, and emergency situation. Of course, the Apollo 13's successful failure was an extreme situation, but it shows that well-organized knowledgeable people with appropriate technology can solve problems in extreme situations. Apollo program people needed a holistic view of the overall endeavor. They did not have the kind of modeling and simulation technology that we have today, but they managed to model and simulate what could happen during an Earth–Moon mission. They had several engineering backgrounds that they combined to make a tangible program. Since then, engineering-oriented human-centered disciplines have evolved toward HSI, together with the evolution of complexity science and artificial intelligence (AI) useful in engineering design.

A holistic view

HSI departs from the traditional approach that considers engineering first and people second, leading to concepts of user interface and training, developed in sequence once an engineered system is almost developed (Figure 0.1).

HSI is holistic, considering a multi-agent world composed of humans and machines. We will talk about human systems and machine systems (the concept of system will be further defined). We now develop a system of systems interconnected via integration links enabling interactivity

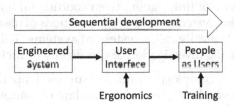

Figure 0.1 Sequential development of an engineered system, then user interface, and finally people's training.

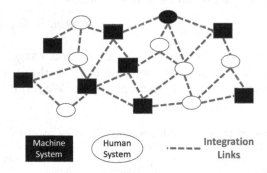

Figure 0.2 Holistic development of a human–machine system of systems.

(Figure 0.2). Having a multi-agent approach does not mean that we should not take care of user interfaces. Most important is to better investigate interconnectivity and coordination among the various agents (i.e., human and/or machine systems) and then derive interfaces among them. We will take the Orchestra metaphor to model such multi-agent systems. In other words, machines are no longer tools but partners.

So, what is the concept covered by this book? The central concept is **engineering design that considers people and organizations seriously during the whole life cycle of a system**. The term "design" can be misleading. In French, we have two words for design: *design* and *conception*. For most of us, the former is about shapes, forms, and esthetics... fashion design for example! The latter is about creation. I will keep "design" for the creation of structures and associated functions, making systems. In other words, **design is about rationalization, modeling, and construction of functional artifacts**.

Virtual human-centered design of complex systems

However, HCD can be used for various kinds of products. In this book, we will focus on **complex systems**, such as aircraft, space systems, oil and gas platforms, hospitals, and defense systems. Complexity is usually coming from the number of elements and links among these elements. It also comes from the type of links, going from conditional branching to feedback loops. Complex systems are also combinations of nested sub-systems, forming what we typically call a **system of systems**. This book offers an approach to attack the investigation of complexity of such systems.

> [What is complex] cannot be summed up in the word 'complexity,' reduced to a law of complexity, reduced to the idea of complexity. Complexity cannot be something that is defined in a simple way and would take the place of simplicity. Complexity expresses a problem and not a solution.[2]
>
> *(Morin, 1990, p.10).*

Human-centeredness is related to operations and usages. HCD breaks with traditional technology-centered engineering that considers operations too late in systems development processes. We should acknowledge

---

[2] Original statement by Edgar Morin in French: *"[Ce qui est complexe] ne peut se résumer dans le mot de complexité, se ramener à une loi de complexité, se réduire à l'idée de complexité. La complexité ne saurait être quelque chose qui se définirait de façon simple et prendrait la place de la simplicité. La complexité est un mot problème et non un mot solution."*

that HCD was not really possible only two decades ago. Conversely, HCD is now an engineering practice in its own right because we have modeling and simulation capabilities that enable human-in-the-loop tests generating very useful guidance for successful HSI, from the beginning of the design process and all along the life cycle of systems being developed and further used. As a matter of fact, HSI can be a goal or a product but also a process that we will call HCD in this book. Consequently, HCD associated to SE leads, or should lead, to HSI.

HCD is now possible because we have **digital support that enables realistic modeling and simulation**. We can develop digital prototypes, also called virtual prototypes, that can be tested very early and contribute to the discovery of emerging behaviors and properties of systems being developed. This is fine and new in engineering design. We now talk about **Virtual HCD (VHCD)**. However, results are digital-native and, therefore, not necessarily tangible from two points of view: physical and figurative (or cognitive).

This book will provide a clarification of **what tangibility of complex systems is about in our digital world, both at design and operations times**.

VHCD contributes to the design of virtual prototypes that require tangibility testing and development of tangibility metrics in our growing digital world.

In addition, we also need to break with current engineering approaches that put technology-centered development first and ergonomic testing after, leading to what is commonly called corrective ergonomics (this will be better explained in the next section of this introduction). The **user interface**, as it is meant today, is a concept of the 20th century (i.e., developing a user interface that enables people to use a machine once the core of this machine is fully developed). Instead, a user interface is a component, as all other components, of a system, which should be designed and tested from the beginning together with the other components.

Virtual prototypes are developed including all system components and, therefore, can be tested in **Human-In-The-Loop Simulations (HITLSs)**, which enable us to test functions allocated to people and machines incrementally by using virtual prototypes. During the design and development processes, virtual prototypes become more tangible, both physically and figuratively through several (re)design and formative evaluations. Such an approach is typically called **agile development** (Schwaber, 1997; Sutherland, 2014). Resulting VHCD enables an iterative **participatory design** process.

At this point, it is important to clarify terms used in SE and AI, especially terms such as agent and system, which have very similar meanings. Indeed, even if this book focuses on VHCD and HSI and, therefore, promotes the concept of **system**, the concept of **agent** is also central to the HSI approach. "An agent is anything that can be viewed as perceiving its environment through sensors and acting upon that environment through actuators." (Russel & Norvig, 2010). The term "agent" is used in AI almost in the same way as the term "system" is used in system engineering. In the same way, a system is a system of systems, an agent is a society of agents. My goal is to promote interdisciplinarity (i.e., AI combined to SE and more specifically VHCD and HSI).

Influences of several engineering backgrounds

Coming from both an engineering background and a cognitive science background, I experienced several frustrations during my career not knowing where I belonged. This is not surprising because the problems I tried to state and solve always required an **interdisciplinary background**, breaking with the verticality imposed by traditionally isolated disciplines. On one side, engineering guided me to design and develop technology by using a positivist approach. On the other side, cognitive science taught me that a human is a whole who cannot be easily split into parts and isolated phenomena. Teared apart between positivism and vitalism (and, therefore, phenomenology),[3] I tried to shape my own HCD approach. We will come back on these two philosophical distinctions later in the book to better understand how they could be combined or not.

More pragmatically, let me explain what triggered the genesis of the concept of **Tangible Interactive Systems** (**TIS**: Boy, 2016) that put forward a combination of (digital) interactive systems and the tangibility concept, thought as physical and figurative (or cognitive) tangibility. Figure 0.3 presents a possible genesis of HSI and TIS from two main engineering backgrounds associated to HCD and SE.

For a long time in engineering, automatic control theories supported automation of mechanical machines, such as autopilots handing various kinds of mechanical parts of an aircraft. The underlying discipline

---

[3] Positivism and vitalism are two different philosophies. The former assumes that the world exists and can be dichotomized into pieces that can be studied in isolation and put back together to understand the whole. The latter assumes that the world cannot be separated into pieces and should be considered as interrelated phenomena. It promotes holistic approaches. These philosophical approaches will be further discussed later in this book.

is mostly based on differential equations and optimization techniques that handle continuous processes. Mechanical and computer engineers progressively came to the definition of embedded systems (e.g., onboard aircraft). They also developed SE. This field recently led to the concept of **cyber-physical systems (CPSs)**.[4]

Studying human factors in commercial aircraft cockpits during the 1980s, I got interested in AI techniques and tools that enable simulation of pilot–aircraft interaction. In practice, I started using automatic control concepts and tools and quickly shifted toward computer science and software engineering support to model and simulate multi-agent complex systems such as an aircrew (two pilots), aircraft, and ground control centers. Shifting from single-agent control equations to multi-agent AI programming techniques was a major endeavor. The shift was **from numerical signals and information to symbolic knowledge representations**. We incrementally developed and piled up software-intensive systems that were more interconnected. More recently, this field mixing sensors and AI led to the concept of **Internet of Things (IoT)**.[5] In addition, computer engineering heavily developed modeling and simulation both for structural parts and functional parts of complex systems.

It is interesting to notice that nowadays CPSs and IoTs are very close conceptually,[6] even if they come from very different backgrounds. In practice, most conferences focusing on CPS are dominated by scientists and practitioners coming from an automatic control background. In fact, CPSs are generally presented as extensions of embedded systems that were developed on aircraft for example. Their foundations are basically assembled electronic and mechanical components that are glued into layers of software. The other way around, IoT is made of computing programs (i.e., software) that are connected to sensors,

---

[4] US National Science Foundation uses the following definition: "Cyber-physical systems (CPS) are engineered systems that are built from, and depend upon, the seamless integration of computational algorithms and physical components." CPS Public Working Group (NIST) provided the definition: "Cyber-physical systems (CPS) are smart systems that include engineered interacting networks of physical and computational components."

[5] Definition 2015 of l'ISO/IEC JTC1 is: "An infrastructure of interconnected objects, people, systems and information resources together with intelligent services to allow them to process information of the physical and the virtual world and react." A IEEE-SA IoT Ecosystem Study produced the following definition in 2015: "IoT refers to any systems of interconnected people, physical objects, and IT platforms, as well as any technology to better build, operate, and manage the physical world via pervasive data collection, smart networking, predictive analytics, and deep optimization."

[6] PICASSO Project Opportunity Report stated that: "In such technical systems, which are often called cyber-physical systems (CPS), real-time computing elements and physical systems interact tightly. ...The merging of IoT and CPS into closed-loop, real-time IoT-enabled cyber-physical systems is seen as an important future challenge." (Sonntag et al., 2017).

enabling interaction with the real physical world (e.g., the Google self-driving car). There is room for a new approach that integrates both backgrounds.

The evolution of engineering-oriented human-centered fields of investigation

Following up on the right-hand side of Figure 0.3 and considering at the same time Figure 0.4, let's describe the evolution of engineering-oriented human-centered fields of investigation for the last seven decades. After World War II and before the eighties, **human factors and ergonomics (HFE)** was the main field of investigation that seriously considered people's issues at work. Before the 1980s, engineering was dominated by hardware; the leading discipline was mechanical engineering. HFE was dominated by physicians (i.e., work medical doctors) looking after health and safety issues at work. They were involved in solving problems dealing with physiological and biomechanical issues. Then came the computer era, more specifically with the development of microcomputers, which invaded our workplaces, public places, and homes. HFE started to shift toward cognitive ergonomics. Information overload was responsible for novel issues at work. We shifted from doing to thinking (i.e., from physical ergonomics to cognitive ergonomics). Doing is executing a sequence of mainly physical tasks. Thinking is about deciding if a sequence of tasks should be executed. Thinking can be goal-driven (intentional) or event-driven (reactive). In any case, HFE has been centered on activity, human performance, situation awareness, and decision-making analyses, in physical or/and cognitive ergonomics.

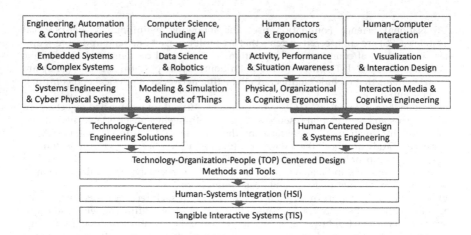

Figure 0.3 A possible genesis of HSI and tangible interactive systems.

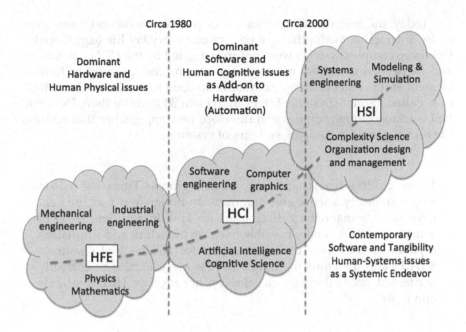

Figure 0.4 Evolution of engineering-oriented human-centered fields of investigation.

Circa 1980s, a new discipline started to emerge in computer science, **human–computer interaction (HCI)**. The first ACM CHI[7] conference was held in Gaithersburg, Maryland, USA in 1982.[8] Since then, CHI conferences have been representative of what can be best in human factors in computing systems.

In Figure 0.4, visualization and interaction design are put forward because they are very significant parts of HCI contributing to HSI, but other HCI components have to be considered, such as text processing, graphical user interfaces, usability engineering, participatory design, and other things related to the user experience of computing systems. People involved are interested in software engineering, user interface design, computer graphics, cognitive psychology, and AI. Cognition has been at the center since the beginning of the 1980s, to the point that it could be said that the real birth of cognitive engineering was in the HCI community (Norman, 1982, 1986), even if other engineering domains produced significant contributions toward the inception of cognitive SE (Rasmussen, 1983, 1986; Hollnagel & Woods, 1983).

---

[7] Association for Computing Machinery – Computer Human Interaction conference.
[8] In 1982, CHI was called Human Factors in Computer Systems Conference.

Today, the field of HCI further develop socio-media and, more generally, interaction media that are part of our everyday life (e.g., Google). However, complex systems were not investigated by the HCI community, and we needed to wait for more systemic approaches putting the human element at the center of complex systems to open a new field of investigation, called Human–Systems Integration circa 2000. Since then, INCOSE[9] HSI working group compiles and develops new approaches that address the role of people in complex systems of systems.

> The new shift is from cognition to socio-cognition. This is the reason why complexity science and organization design and management have become mandatory disciplines for HSI. In addition, HCD of complex systems is now possible because modeling and simulation capabilities offer very realistic support for HITLSs that results in **VHCD**, and therefore, the tangibility issue remains to be correctly addressed and solved by developing appropriate HCD methods and tools.

Another way of describing such an evolution from a human–science perspective is presented in Figure 0.5. This view requires us to understand the distinction between task and activity. The **task** is what is prescribed to be done by one (or several) human operator(s) or a machine user(s).

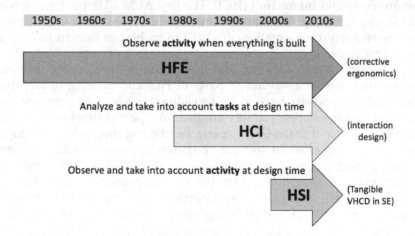

Figure 0.5 Evolution of human-centered fields with respect to task and activity analyses.

[9] International Council on Systems Engineering.

Activity is what is produced effectively by human operators or users. The difference between task and activity of a human agent is usually caused by intrinsic and extrinsic events.

Intrinsic events may be caused by physical, physiological, or psychological processes constraining human performance (e.g., fatigue, stress, workload). **Extrinsic events** may be caused by sociological, organizational, or environmental processes impacting human performance (e.g., interruptions, failures, physical, and/or social disturbances). We can talk about task and activity at various levels of a human–machine system (i.e., agency of agents or system of systems). Indeed, a company (i.e., a system of systems) has usually tasks to be performed and actually performs an activity. Such activity is a combination of activities of individual agents.

Considering the task–activity distinction, HFE, HCI, and HSI can also be distinguished and better understood. **HFE** is really based on **activity analysis** since the beginning – activity theory is coming from social constructivism, developed by Soviet's scientists such as Lev Vygotski, Alexei Leont'ev, and Sergei Rubinstein, later on by the French ergonomics society (SELF[10]), Nordic European Ergonomists, and HCI US scientists (Kozulin, 1986; Amalberti, de Montmollin & Theureau, 1991; Nardi, 1995; Bertelsen & Bødker, 2003; Leplat, 2008; Kaptelinin & Nardi, 2006). During the 20th century, activity analysis was only possible before design (i.e., analysis of human operator's activity within the scope of an old existing system) and after system manufacturing (i.e., activity analysis using the new fully developed system). Difficulty with this approach was that activity analysis based on the use of an old existing system could not anticipate what would emerge with the use of a new system, and activity analysis using the new fully developed system turns out to be often too late to make deep modifications when necessary. This is the reason why user interfaces and operations procedures are still developed to compensate some design flaws with respect to HSI. This is the reason why we usually talk about "**corrective ergonomics.**"

Starting during the early eighties, **HCI** was developed by computer scientists and cognitive scientists to provide people with useful and usable computing systems. They developed new approaches to task analysis because in computing applications, such as text processing, computer graphics, speech recognition, and other processes that required computer programs to support their tasks, task and activity turn out to be very similar (i.e., very little discrepancy exists between task and activity in this kind of HCI). HCI scientists and practitioners developed **interaction design**. Unlike the HFE community, HCI community was never limited to evaluation and developed design methods and supporting tools since its inception.

---

[10] Société d'Ergonomie de Langue Française.

Development of complex systems now starts by developing virtual prototypes that enable HITLS, which supports activity observation and analysis. Most importantly, HITLS enables discovery of emerging behaviors and properties of systems being incrementally designed and tested. **HSI in SE** of complex systems is made possible by **VHCD** associated with a tangibility-based approach. As a matter of fact, VHCD is mostly used to test functionalities (i.e., human and machine function allocation). Main problem with this very innovative approach, which considers human factors effectively, is **tangibility**. Of course, HSI in SE encapsulates conventional HFE and HCI.

HCD supported by virtual prototypes (i.e., VHCD) requires the following capabilities:

1. Good cognitive engineering background
2. Complexity analysis, organizational design, and management
3. HITLS practice at design time, supported by appropriate modeling and simulation tools
4. Tangibility exploration and assessment.

It is important to note that complex systems that we are talking about in this book are **life-critical** (i.e., these systems involve safety, security, efficiency, and comfort issues). You may ask yourself why include efficiency and comfort into life-criticality. For a long time, I was interested in safety-critical systems, and the question was clear (i.e., only a single answer: "safety"). However, consider the smartphone that you use every day (if you use one of course!). If you lose it, you will start to understand that this is a life-critical system, because your life will suddenly become less efficient (e.g., you will not be able to call for anything critical) and comfortable (e.g., consult your agenda to help you remember the next appointment or meeting).

> To summarize at this point, HSI can be described as both a process and a product. As a process, HSI involves associating HCD with SE. As a product, HSI is the result of a good integration of TOP. HSI is the intersection of several disciplines such as ergonomics, SE, and computer science. We will see later in the book that HSI should also consider economic constraints and environmental issues.

The evolution of complexity science

This book is mainly about systems of systems, which emphasize systems that are necessarily interconnected (i.e., not isolated). Complexity mainly comes from interconnectivity that includes numerous and/or feedback loops. This is the reason why we need to know about complexity science.[11] This section briefly presents the evolution of complexity science in order to better understand where we stand now regarding this important topic.

General systems theory (GST) was founded by the Austrian biologist Ludwig von Bertalanffy, as an interdisciplinary approach describing what we call today a system of systems (Bertalanffy, 1968). Bertalanffy developed a mathematical model representing growth of a living organism over time. His model turned out to applicable to biology, cybernetics, and information theory. As a matter of fact, GST was close to Norbert Wiener's cybernetics[12] (Wiener, 1948), and both of them led to **complexity science**.[13]

The concept of **emergence** is crucial in a complex system (i.e., apparition of new characteristics when the system is at work). Emergence cannot be clearly understood without understanding the concept of "phenomenon." A phenomenon is a manifestation or appearance of something. It comes from experience. We find phenomena in physics, biology, ecology, socioeconomics, linguistics, and other dynamic systems that include retroactions.[14] Emergence cannot be clearly understood without understanding that a system is more than the sum of its parts. In other words, we cannot predict the overall behavior of a complex system from the analysis of its parts independently. Emergent properties contribute to the evolution of organizations, initiating self-organization.

Why should we take care of complexity in HSI? First, systems that we develop are getting more interconnected every day. Software is everywhere, which means that local systems promote individualization (i.e., systems can provide more autonomy to people and local

---

[11] For example, engineers are good at simplifying a problem to quickly find a solution. However, when simplification is decomposing (i.e., separating) a complex system into parts, it is important to know about the separability property in systems of systems (i.e., a sub-system can be studied separately from the rest of the overall system). We will further develop separability in the following.

[12] Wiener, Rosenblueth, and Bigelow contributed to define what control theory is about today, by challenging behaviorist orthodoxy and promoting regulatory mechanisms that minimize error between current state and goal (Rosenblueth et al., 1943). More generally, Bertalanffy was interested in representing open systems. He argued that open system phenomena cannot be explained by the second law of thermodynamics, which assumes that the entropy of an isolated system can never decrease over time.

[13] I recommend Melanie Mitchell's book that provides a good introduction to complexity science (Mitchel, 2009).

[14] The retroaction concept is directly related to homeostatic systems that find their roots in Claude Bernard's work at the end of the 19th century. Bernard, a French physiologist (1813–1878), promoted what is called the *conservation du milieu intérieur* (preservation of system's internal structures and functions). This preliminary work led to several pieces of work by the American physiologist Walter Cannon (1871–1945), the American mathematician Norbert Wiener (1894–1964), and the English psychiatrist Ross Ashby (1903–1972). Ashby created a homeostatic device, including feedback loops, which were able to achieve adaptive behavior and, therefore, a kind of artificial intelligence (Ashby, 1952).

organizations). However, the more local systems become **autonomous**, the more they need **coordination** rules to keep the integrity of the overall system of systems.

> An autonomous system should be able to independently make decisions among different courses of action to accomplish goals based on its own knowledge, situation awareness, and purpose.

Complexity should be investigated because we need to be aware and consider emergence of new properties. Consequently, we need to know how emergence works and can be observed, assessed, rationalized, and projected. In other words, we need to rationalize what TOP involved induce. Moreover, our global economy increases the number of "everything" (e.g., recorded and processed data), which increases complexity everywhere in systems that we are developing and using. In addition, complexity comes from the lack of understanding of what is going on in the background (i.e., lack of figurative tangibility and situation awareness). Therefore, we need to look for new TOP[15] solutions.

Four kinds of complexity properties can be considered:

1. Principle of **attraction**, mathematically modeled by strange attractors in chaos theory, for example (e.g., desires for power, for love, pleasures, freedom)
2. Principle of **fractality**[16] that covers similar structures that repeat at different scales of generalization (e.g., individual system dynamics nested in family system dynamics)
3. Principle of **emergence** of new dynamic patterns (e.g., looking at life evolution, several life forms emerged, inorganic matter gave birth to organic, and from the organic matter the realms of plants, animals, and humans emerged)
4. Principle of **self-organization**, which involves *self-organizing forces* necessary to sustain emergence, without such forces, no one emergent form can survive (e.g., interacting moving air streams with different physical or/and chemical characteristics drive tornado's spiral turbulence).

---

[15] Technology, Organizations, and People.
[16] The French–American mathematician Benoit Mandelbrot (1924–2010) studied a mathematical object that he called a fractal, which is a shape where each part is like the whole object but smaller. Fractals have been observed in nature (e.g., the cauliflower, the human lung). This representation that expand and unfold symmetry can be very interesting when we will study a system of systems. This replication property is called self-similarity.

We often say that to avoid developing complex systems, we need to simplify! **Simplification** is a reductionist approach that usually leads to complexity issues such as catastrophes, in the sense of René Thom.[17] For example, when we project a 3D space onto a 2D space, we simplify but we also create ambiguities due to the elimination of the third dimension. Instead, it is often better to get familiar with the complexity of an object in order to understand its properties. In HSI, I would often privilege the effort of **familiarity**[18] to the process of simplification... it depends of course on the case! Familiarity can be facilitated by developing a model of the system we want to investigate. It is difficult and often impossible to measure anything without a model.

Managing complexity by designing and developing a phenomenological model

This is the reason why the model will help defining data elicitation or acquisition from real-world phenomena (Figure 0.6). Modeling is a hypothesis-driven approach to complexity. It is also a projection into the future that involves an abductive inference (i.e., we will need to demonstrate that our projection is correct). Modeling is making an abstraction. It is practicing figurative tangibility (this concept will be further explained later in this book).

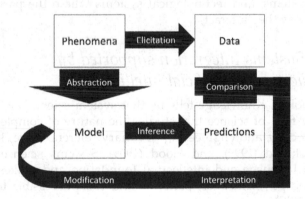

Figure 0.6 Modeling process that increases familiarity with complexity of phenomena.

[17] The French mathematician René Thom (1923–2002) studied topology and singularity theory. He founded and developed catastrophe theory and received the Fields Medal in 1958.

[18] Familiarity will be further defined in Chapter 4. At this point, let's say that best familiarity's synonyms are acquaintance, awareness, experience, insight, comfort, naturalness, and simplicity.

The model enables inferring possible predictions that can be compared to experimental data measured on real-world phenomena. Such comparison, going from data to information to knowledge, enables interpretation and further modification of the model. Several iterations are often needed to validate the model, as well as its context of validity.

Complexity science requires rich background of theoretic work. It addresses deep scientific questions about nature. In our case, we are building systems that are not natural, but because they have become so interconnected, they look like **living systems.** It is not the same to design and build a fork or a knife as to design and develop an air traffic management (ATM) system. ATM design requires us to know how to design, manage, build, and control systems as they tremendously increase in size and connectivity. Up to a point, **size** (i.e., number of elements and connections among these elements) matters so much that emerging properties pop up and require to be considered seriously. Therefore, we need to build systems that are scalable, robust, and adaptive by using properties such as self-organization, self-adaptation, and self-repair that almost only characterized biological systems up to now.

Complexity science is a broad and multidisciplinary field that supports the scientific study of complex systems. Examples of complex systems are IT networks, ATM systems, ecosystems, brains, markets, cities, and businesses. Santa Fe Institute[19] contributes to production of knowledge in the field (e.g., sand piles, complex adaptive systems such as social systems, biological systems, and technological systems where the parts actively change the way they interact).

Human–Systems Integration supported by system science and artificial intelligence

INCOSE Book of Knowledge tells us that system science[20] is "an interdisciplinary field of science that studies the nature of complex systems in nature, society, and engineering." Primary references are Bertalanffy (1968), Checkland (1999), and Flood (1999). Systems science provides "a common language and intellectual foundation and makes practical systems concepts, principles, patterns, and tools accessible to systems engineering (SE)."

During the 1980s, AI developed so much that we were thinking that it could invade our lives and replace people. Almost three decades followed with an AI winter! Today, AI resurrects even bigger that before. Should we be worried about being replaced by machines? Or, should we think in terms of interacting and collaborating with smart machines?

---

[19] www.santafe.edu
[20] www.sebokwiki.org/wiki/Systems_Science

The Cloud, for example, brings more autonomy to people than any tool had provided before. However, we need to be cautious. We should look after AI algorithms' flexibility and **maturity**. We should make sure that AI does not bring ways of doing things that are more complicated. Think about voice menus when you call a large company; you usually end up in being extremely frustrated just because the system is too rigid. This is because such voice recognition systems were immature when they were developed and were never modified.

Sometimes, AI should not be based on human cognition only but on other forms of intelligence when it makes sense to do so. Look at a flock of birds. Isn't it smart? Thousands of birds flying close to each other creating a majestic pattern. This is natural **collective intelligence**. Look at the rise and fall of species in evolution (i.e., species that were, or were not, capable of adapting to the evolution of their environment). Look at interactions of people and groups in a community. Interactions in social groups, teams, communities, and organizations have their own intelligence that is interesting to be modeled and further understood. This is the reason why we need to have a good educated common sense of systems whether they are natural or artificial. This is not possible without experience. This book will open a new way of considering systems for HSI, not only by considering human cognition but also natural life. More generally, a system-of-systems approach will be adopted in this book. In addition, we need to have a framework that encapsulates both natural and artificial systems, where **information** and **energy** can be both expressed independently and combined to explain complex behaviors.

Finally, numerous complex problems are not solved the way the Apollo 13's successful failure was. We need to remember this kind of proactive engagement of a team of experts in charge of an extremely difficult issue. For example, current climate change issues should be thought and processed using this kind of approach instead of focusing on isolated disciplines and not on interdisciplinary problem solving. The human element should definitely put at the center of our major sociotechnical problems. We should consider creativity, unexpected events, and collaborative work seriously.

A systemic human-centered design approach toward Human–Systems Integration

The book is structured around this introduction, nine chapters, and a conclusion.

Chapter 1 presents the human element in industrial complex systems from a digital world perspective, more specifically the evolution from technology-centered engineering to HCD. VHCD processes will be presented as design documentation management, virtual prototyping, digital twins design and use, HITLS support and practice, agile development, and formative evaluations.

Chapter 2 introduces tangibility problems and potential solutions. It explains the engineering shift from 20th to 21st century and defines tangibility from physical and figurative viewpoints. Attributes, models, and metrics are provided for tangibility evaluation. A method for incremental construction of tangible things is presented based on the development of conceptual models (i.e., ontology).

Chapter 3 presents the various assets of TOP polarized into two models: the TOP model and the AUTOS pyramid. Societal and cultural issues are discussed, as well as impact of new technology on organizations and jobs. The evolution of engineering design from means-driven to purpose-driven is developed. A special focus on the articulation of structures and functions is provided.

Chapter 4 focuses on life-cycled HSI (i.e., HSI from design, manufacturing, certification, delivery, training, operations to decommissioning). System is presented as a representation of machines and people. A system has at least one structure and one function. It can be cognitive and/or physical. A system is in fact a system of systems, leading design to a multi-agent approach. HSI cannot be meaningful without being contextualized. Therefore, the concept of context will be further developed. Finally, two models, natural/artificial versus cognitive/physical (NAIR) and structure/function versus abstract/concrete (SFAC), will be provided to integrate physics, cognition, philosophy, and design.

Chapter 5 is devoted to the shift from rigid automation to flexible autonomy. The evolution of automation using Rasmussen's model, linking human behavioral processes to machine equivalent processes. Designing for flexibility will be promoted based on the TOP model, and a few examples will be provided in normal, abnormal, and emergency situations. Looking for flexibility is a matter of function analysis and allocation. The emergence of AI and HCI will be explained as major technological domains for automation design and management of complex systems. It will be shown that rigid automation is mainly a matter of maturity and costs.

Chapter 6 introduces orchestration of HSI using VHCD. It presents what it takes to model and simulate in a human-centered way using computer-aided design software technology leading to virtual prototypes. VHCD will be presented as an agile development method that involves formative evaluations and iterative redesign. The first phase is knowledge elicitation from domain experts. The Group Elicitation Method will be presented, associated with design thinking, and creativity methods. The second phase consists in carrying out HITLS, which enables the design team to be more familiar with complexity of the system being designed and developed as well as its use. I promote educated common sense, structure–function symbiosis, and minimality.

Chapter 7 focuses on design for flexibility in an increasingly complex world. The reductionist issue in engineering is discussed leading to risk analysis and complexity of unexpected events, which themselves induce problem solving, by default of procedure following and appropriate automation. Several categories of life-critical systems will be described with respect to risk they induce. Complexity analysis benefits from the separability property (i.e., some parts of a complex system can be separated from the others, and some are not separable, i.e., they should be investigated connected with the other parts), as well as HITLS that enables to discover emerging behaviors and properties. Associating risk with complexity, we will look for attractors, rationalizing normal, abnormal, and emergency situations. We will see how designing for flexibility in such situations requires appropriate socio-cognitive representations. System-of-systems flexibility and autonomy will be presented. More specifically, we will discuss the issues of authority sharing and accountability. A model of resource and context management for HSI will be introduced.

Chapter 8 is devoted to storytelling, creativity, and tangibility toward innovation. Storytelling is about soliciting experts and experience people to explain what they usually do. Such explanation requires methods such as timeline analysis, activity analysis, and journey mapping. Design thinking and creativity is about problem stating, problem solving, and solution delivery. Some methods will be introduced. Divergent and convergent thinking should be associated for successful and tangible results.

Chapter 9 is about evaluation and metrics of tangibility within the framework of VHCD. Several concepts will be presented and discussed, such as quantitative versus qualitative metrics, evaluation context and operational context, and a method for the evaluation of a system of systems. Various high-level evaluation approaches and metrics will be presented within the AUTOS pyramid design framework.

In conclusion, perspectives will be provided that include research needs, leveraging HSI above current industrial constraints (essentially due to our economic system worldwide). We will discuss the evolution of AI from symbolic reasoning and machine learning, which involves longer-term investigation of possible futures, to data-driven induction and deep learning, which involves short-term predictions. AI is certainly the science that will enable design of increasingly autonomous systems; therefore, it is crucial to understand the right direction we should take (i.e., anticipating, understanding and evaluating possible futures, or accurately predicting in the very short term).

chapter one

Human-centered design of industrial complex systems

> That's what we do in real life, with puzzles that seem
> very hard. It's much the same for shattered pots as
> for the cogs of great machines. Until you've seen
> some of the rest, you can't make sense of any part.
>
> **Marvin Minsky**
> *Cognitive Scientist and Artificial Intelligence Pioneer*

There are three main difficulties that arise when humans want to understand and interact with a complex system. They need to know (1) how the whole system works, (2) why the system exists, and (3) how it should be used. During the 20th century, even if each part was very well tested independently, we needed to wait until the overall system was assembled (i.e., integrated) to test it as a whole. Today, we can test an entire complex system with a human-centered design (HCD) approach using virtual prototypes and carrying out Human-In-The-Loop Simulations (HITLSs). This is what we call Virtual HCD (VHCD).

Technology-centered engineering versus human-centered design

Software engineering proposed the V-Model (Figure 1.1) as an extension of the waterfall model. It is currently used by most industrial companies worldwide, not only in software engineering, but in any kind of engineering. It provides a graphical "picture" of a development life cycle of a system. The left side of the "V" shows the sequence of processes going from requirement analysis, to system design, architecture design, module design, and coding of the parts (i.e., manufacturing in the general case). The right side of the "V" shows integration of parts, through unit testing, integration testing, system testing, and finally acceptance testing (certification).

We observe that most difficulties encountered in the validation phase come from requirements that were not enough human-centered or organization-centered. In fact, we pay high price in the end on lack of sometime-small requirements flaws. We end up with a "V," directed by technology-centered engineering (TCE), that looks like a check mark

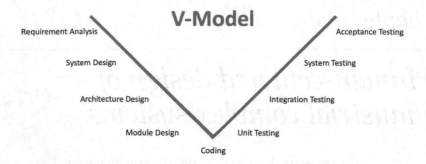

Figure 1.1 V-Model.

(Figure 1.2), that is, we spend extra amount of time and pay a lot of money to be ready before delivery (when it is not too late to still fix design flaws). Obviously, TCE is currently combined with human factors and ergonomics (HFE) considering that HFE is mobilized at the end of the V-Model for acceptance testing.

However, requirements need to be validated first against the higher-level requirements or user needs (i.e., defining human-centered requirements). This is when modeling and simulation (M&S) comes in, and an HCD contribution can be developed. As a matter of fact, HCD can be symbolically represented as an exact symmetric shape of TCE check-mark V-Model (Figure 1.3). Note that if HCD obviously includes HFE for the design parts, TCE also includes HFE for the testing parts.

We already claimed that combining HCD and TCE leads to Human–Systems Integration (HSI). In addition, the amount of effort required without HCD can be reduced carrying out HCD processes. Finally, the more we get experience carrying out HCD processes, the more the amount of effort required for TCE will decrease (Figure 1.4).

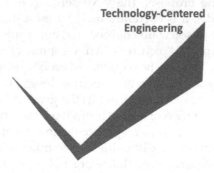

Figure 1.2 The V-Model when not enough attention is brought to human-centered requirements (the "qualitative thickness" of the line expresses the amount of effort and costs).

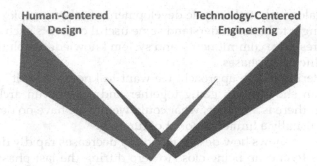

Figure 1.3 Combining HCD and TCE.

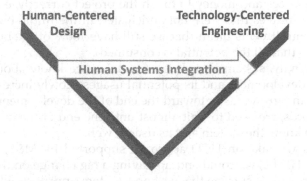

Figure 1.4 Combining HCD and TCE to develop HSI.

We will describe later in the book how HFE testing procedures and criteria should be adapted by considering tangibility metrics.

Impact of virtual HCD on the life cycle of a system

It is important at this point to explain what we mean by HCD, and more precisely VHCD. I already provided a description of HCD (Boy, 2013a), which includes six components: cognitive engineering, complexity analysis, M&S, advanced interaction media, organization design and management, and life-critical systems. M&S enables building virtual prototypes that are key HCD digital resources based on artificial intelligence (AI), human–computer interaction, and data science. It is now possible to immerse, very early in the design process, potential users, human operators, and simply people in the loop with simulated systems prefiguring targeted systems to be developed.

We will take NASA's life cycle phase model (Table 1.1) to illustrate what M&S, and therefore VHCD, can provide to improve the life cycle of a system. This model is obviously specific to space systems, but it can be applied analogously to life cycle development of other complex systems. We are using it to better understand some useful concepts such as design flexibility, resource commitments, and system knowledge evolutions with respect to life cycle phases.

Considering a TCE approach, we want to know what "it" will look like. We can show pieces going together and develop an architecture. However, if there is a change, what could we do; we have no design flexibility and usually a limited amount of funding.

Figure 1.5[1] shows how design flexibility decreases rapidly during the first phases to end up being close to zero during the last phases. In the same way, resource commitments increase rapidly during the first phases to end up on a saturation during the last phases (i.e., we will not have enough resources and money to finish the project correctly, except if we add more resources that were not anticipated in the initial budget – 0% resource commitments mean that we still have 100% of the budget, and 100% means that all the potential is consumed).

Interestingly, system knowledge (i.e., what we know about the system under development and its potential usages) slowly increases in the beginning and grows faster toward the end of the development life cycle. In other words, we need to wait almost until the end of the development life cycle to know the system and its usages well.

Now, if we take an HCD approach supported by M&S, and more specifically HITLS, we could end up having a big change on these evolutions (Figure 1.6). Assuming that we know system's purpose, when people ask, "What will the system look like and how it could be used?", we could show them a simulation and even get them involved into the simulation.

Table 1.1 NASA's life cycle phases

Pre-Phase A, Concept Studies	Feasible concepts, simulations, studies, models, mockups
Phase A, Concept and Technology Development	Concept definition, simulations, analysis, models, trades
Phase B, Preliminary Design and Technology Completion	Mockups, study results, specifications, interfaces, prototypes
Phase C, Final Design and Fabrication	Detailed designs, fabrication, software development
Phase D, System Assembly, Integration and Test, Launch	Operation-ready system with related enabling products
Phases E and F	Operations and sustainment, closeout

[1] I thank Mike Conroy for the work he supported when I was at NASA Kennedy Space Center, as well as his involvement into INCOSE HSI Working Group.

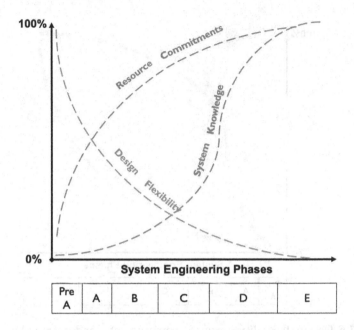

Figure 1.5 Design flexibility, resource commitments, and system knowledge evolutions (Conroy, 2016) with respect to life cycle phases (e.g., NASA's phases) using conventional TCE.

Most interesting is that system knowledge is increasing much faster in the beginning because M&S enables to observe various kinds of activity as well as emerging behaviors and properties. In addition, M&S helps the design team in finding out alternative design solutions; we keep design flexibility for a longer time during the design life cycle of the project. Resource commitments start to grow slower with M&S support, and we have more time to make "final" design decisions. At this point, resource commitments are higher in the beginning due to investment in M&S, but we could reasonably expect that initial M&S expenses can be cushioned over a short period of time considering advantages due to improvements in system knowledge and flexibility.

Of course, this is a pedagogical exercise. Indeed, it can be anticipated that phases will not be the same in duration between Figures 1.5 and 1.6. In particular, we could expect that initial project phases might be longer and final phases shorter. All this depends on budget, system complexity, personnel expertise and experience, and organizational setup.

As a business case, let's assume that we want to design and develop a truck platooning system (i.e., linking two or more trucks among each other in a convoy). The goal of this system is to automatically keep a close distance

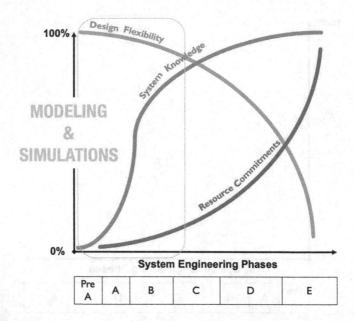

Figure 1.6 Design flexibility, resource commitments, and system knowledge evolutions (Conroy, 2016) with respect to life cycle phases (e.g., NASA's phases) using VHCD (i.e., M&S).

between two interconnected trucks and manage a "train" of trucks on a road (e.g., motorway). We need to define conditions, advantages, and drawbacks for safe truck platooning. HCD dictates that we need to study not only truck driver's activity but also the activity of other drivers involved in the traffic that includes truck platooning. Such human activity can be examined considering performance, workload, attention, vigilance, stress, fatigue, situation awareness (SA), human–machine cooperation, fatigue, and comfort, for example. We also need to investigate the impact of platooning on traffic fluidity, road safety, environment, and socioeconomics.

We then need to organize an M&S process that will increase system knowledge, enable incremental definition of resources, and guaranty flexibility for later changes due to the evolution of technology and market constraints. It is then clear that a truck simulator is necessary for HITLS tests enabling the discovery of emerging behaviors and properties (i.e., we need to put real humans in the loop in order to elicit such credible properties of the overall system). We immediately see the need for a multi-agent simulator that enables several drivers to interact among each other both inside (i.e., truck drivers) and outside (i.e., other vehicle drivers) the platooning process. This may seem to be increasing cost in the beginning of the project, but the elicitation of meaningful emerging behaviors and properties will increase system knowledge and, therefore, improve system maturity.

The most important in the evolutions of the three parameters (system knowledge, design flexibility, and resource commitments) versus the life cycle of a system is the consideration of the three at the same time (Figure 1.6).

Virtual human-centered design as document management

During my childhood, I used to go watch the carpenter working in his workshop. I liked the smell of freshly cut wood incrementally transformed into all kinds of things. Everything was tangible. I could smell, touch, grab, and hold this physical matter. Today, I never stop typing on my computer, writing, drawing, and so on. No smell, no physical feeling, no proprioceptive feedback... I am connected to a tool cognitively. Everything I generate is virtual. Of course, writing this book is a matter of providing other people, you, the reader with material for the mind. Some people would say, this is intangible! I hope it is understandable, that is tangible figuratively. In other words, you are able to feel, touch, and grab the concepts that I am developing. It will be pretentious to say that I am designing concepts that you will be using, in the same way the carpenter is designing wood things that other people will be using. In any case, we need to rationalize and make what we design concrete both physically and figuratively.

Writing is design, and design is writing (Norman, 1992). Indeed, writing requirements help rationalize systems to be or being designed. Writing evaluation results of system testing also helps rationalize what went good or wrong, and therefore, propose alternative solutions for redesign. We are talking about documenting the design process and its solution (Boy, 1997b). For a long time, in traditional TCE, engineers were writing requirement documents for manufacturing services who developed systems, which were documented operationally for appropriate usages (Figure 1.7).

This is typically what industry has done for years. It is interesting to notice that good practice in HSI incrementally promotes a feedback loop from operations documents and requirements documents. This is almost impossible considering the sequential technology-centered approach presented in Figure 1.7. A better model is presented in Figure 1.8, where paper-based documentation has been transformed into an electronic documentation system. Designers, manufacturers, legislators and certifiers, and end users are able to interact with the "same" media. As a matter of fact, even if these stakeholders deal with the same system, they usually have to use such an electronic technical documentation (ETD) for different purposes. In other words, such a media should be contextualized.

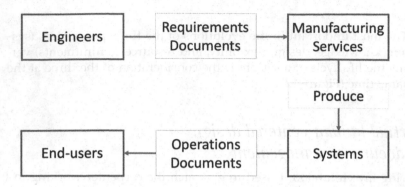

Figure 1.7 Traditional TCE production of industrial documents for manufacturing and operations.

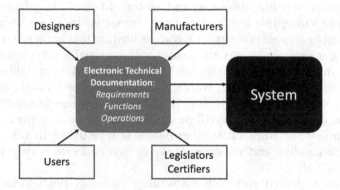

Figure 1.8 ETD interrelating requirements, functions, and operations (e.g., operations procedures and checklists) useful to deal with a system.

This is the reason why the active design document (ADD) representation has been developed and further used (Boy, 1997b).

ADD knowledge representation was initially developed to support traceability of design (and modification of systems) decisions, as well as rationalization of innovative concepts and incremental formative evaluations (Boy, 2005). Considering the task–activity distinction (already presented in this book), an ADD was originally defined by three components: a task space describing how the system should be used (e.g., procedures); an activity space offering potential user capabilities of using the system (e.g., simulator); and an evaluation space offering stakeholders dealing with the system (e.g., designers, evaluators), capabilities of creating new entries, and modifying existing ones (e.g., usability test results, requirement modification suggestions).

Later on, the **design card** (DC) representation was introduced (Boy, 2016). DC supersedes ADD. Like in the ADD, DC includes a rationalization space and an activity space. The activity space provides dynamic visualization capabilities that enable the manipulation of systems being designed and developed. Instead of ADD's task space and the evaluation space, DCs have a structure space and a function space. While ADDs emphasize HFE issues only, DCs emphasize HSI as a whole concept. This is why structure and function are put to the front (see Chapter 4). Structure space emphasizes the multi-agent declarative perspective (i.e., the system-of-systems view). Function space emphasizes allocation of both physical and cognitive functions and supports storytelling (i.e., scenario-based design) from the start of the design process.

A DC is defined by four entities (Figure 1.9):

- A rationalization space where the various components of the system being designed (SBD) are described in terms of design rationale, integration, and requirements; this space includes declarative and procedural descriptions and statements.
- An activity space where the current version of the SBD is displayed; this space includes static and dynamic features, and enables SBD manipulation.
- A structure space where the various components and their interrelations are formally and declaratively described as systems of systems.
- A function space where the various functions of the SBD are described in terms of procedural knowledge and dynamic processes involved; this space includes qualitative and quantitative physical and cognitive models.

A given DC presents the state of the design of a system at a given time for a given design team member (DTM). It is formally represented by DC (t, DTM_i), where t is the time and DTM_i is the design team member i (could be a person or a group of persons).

A DC enables designers to describe the various components of the whole system in the rationalization space, display and manipulate them in the activity space, describe and use the navigation and control features in the operational space, and fill in the evaluation space as required after

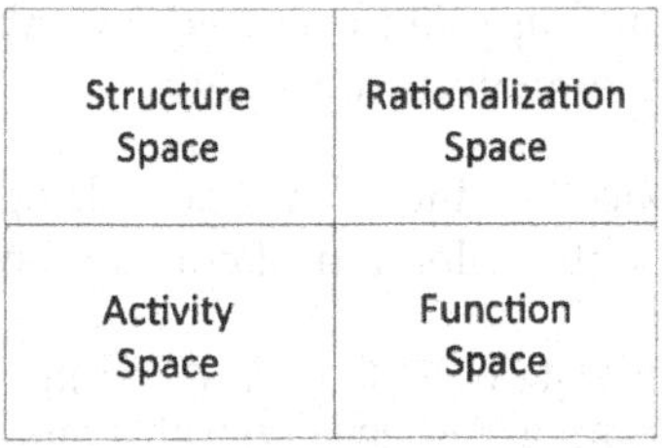

Structure Space	Rationalization Space
Activity Space	Function Space

Figure 1.9 Design card (DC).

assessment of the SBD. An example of DC is provided in the I²S-LWR design project.[2]

The upper-left part is the structure space where the system is described in terms of abstract concepts and their interrelations. The lower-left part is the activity space where the system and/or its components and dependencies can be visualized and manipulated as virtual objects. The upper-right part is the rationalization space where design rationale can be stored and related to the three other parts. The lower-right part is the function space where physical and cognitive functions can be defined, refined, and connected to the three other parts. In addition, any DTM can interact with another DTM using the instant messaging space of the DC. All DC parts are interrelated. For example, a DC user can easily navigate from one part to another. The current version is in progress. Later versions will enable any DTM to generate any component and describe it in the various spaces.

Using DC supports solving several problems such as geographical spread-out of experts of these groups, speed of technology evolution, high personnel turnover, and lack of documentation of the design process. DC generation happens during design. When DCs are documented regularly, they do absorb very little time of the design process. This additional time is compensated by a gain of time due to shared situation awareness (SSA) of the entire design team. DC quality contributes to the quality of design.

Each DC (t, DTM_i) corresponds to a version of the system being designed and developed. Each time design management has a design review meeting at time t_1 (Figure 1.10), all DTMs analyze the work done by each DTM and create a synthetic ADD (t_1, DT), where DT is the whole design team. DCs are like scores that musicians use to play a symphony in an orchestra, with the peculiar difference that, unlike scores, DCs are being incrementally defined to get a sound symphony in the end of the design process.

After a design review meeting at time t_1, each DTM_i works on the premises of DC (t_1, DT) and produces their own DC (t, DTM_i) until the next design review meeting is organized at time t_2, where a new DC (t_2, DT) will be produced from the integration of all ADDs created and/or modified by each DTM during the time interval $[t_1, t_2]$.

Each DC is stored into a design database and can be retrieved at any time by any member of the design team (although some restrictions could be implemented and applied if necessary). Various DC traceability mechanisms can be implemented such as via:

- Rationalization space (by date of creation, design rationale keywords)
- Visualization space (by selection of components graphically)

[2] The Integral Inherently Safe Light Water Reactor (I²S-LWR) nuclear project is funded by the U.S. Department of Energy, in collaboration with several universities and industry partners including Georgia Tech, University of Michigan, Florida Institute of Technology and Westinghouse (Boy et al., 2016).

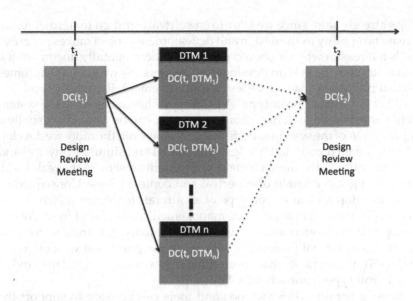

Figure 1.10 DC's generation.

- Operational space (as a table of contents)
- Evaluation space (using scenarios and criteria as indices).

Modeling and simulation: virtual prototypes and digital enterprise

ETD can be considered nowadays as an interactive multimedia system. Taking back the double association "writing as design" and "design as writing," writing an ETD helps designing the system that it documents, and designing a system can be done by tangibilizing[3] the related ETD. In the beginning of the design process, the first ETD can be a shallow user interface showing the various stakeholders what the final product would be (e.g., in the same way an architect would show you your future house). Incremental tangibilization of ETD leads to the concept shown in Figure 1.11.

- Writing an ETD is VHCD of the related system.
- Reading an ETD is interacting with the related system via this ETD.
- Reviewing an ETD is testing usability of the system it documents.

---

[3] The terms "tangibilizing," "tangibilize," and "tangibilization" are neologisms, coined in this book to express action of making things more tangible.

As already said, since we start from software and go to hardware, we increase tangibility as the design and development project makes progress. This is the reason why we should use DCs to incrementally document the various steps of the system development process. As we go, ETD becomes a virtual prototype that can be tested and incrementally redesigned.

What is a **virtual prototype**? A prototype anticipates what the system it represents should look like, should work and can be used. It is usually a simplification of the system to be developed. However, the more we develop and collect data (i.e., digital "things" that model and simulate physical and cognitive entities), the more virtual environments resemble the real world, at least in a given domain of expertise and context of use. Consequently, we can develop a virtual prototype of an aircraft for example, which can be connected to a virtual air traffic management system, and fly it! We can develop virtual interconnected systems of systems. We are able to push the realism of virtual prototypes very far to the point that we can generate HITLS environments that are credible real-world analogs. This kind of virtual prototypes supports VHCD.

More generally, M&S methods and tools can be used to support the overall enterprise (i.e., the network of organizations where the system is being designed, developed, manufactured, tested, certified, operated, maintained, and dismantled). **Digital enterprise** is a comprehensive network of digital models and methods, which includes process dynamics simulation and 3D visualization where humans can be involved (i.e., like in a video game). Digital enterprise purpose is for us to consider the overall enterprise; plan holistically; develop the whole thing as a system of systems; and use the resulting digital product as a tool to control, maintain, repair, and even modify both processes and the overall system.

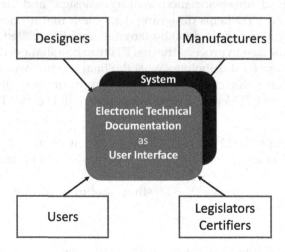

Figure 1.11 Electronic technical documentation as user interface of a system.

Dassault Systèmes and Sogeti developed the Virtual Concept for digital manufacturing and simulation (Coze et al., 2009). They clearly showed the real benefit of using virtual prototypes in industrial engineering design. More specifically, they showed how validation in a virtual world greatly improved, compared to validation in a physical world (only), in terms of time, modification cost, and modification possibilities. Although the Virtual Concept was promoted and is actually consolidated today (10 years after the production of Coze et al.'s book), because it leads to reduced time and cost early in product development, making the case for early simulation in manufacturing, the proper combination of virtual and physical tests holds the potential for increased innovation while reducing costs and time to market. This means that VHCD is great for innovating by functionally testing new features, but it requires tangibility testing anyway (requiring real-world flight tests in aeronautics, for example). The Virtual Concept is easily expandable to maintenance.

Digital twins

A digital twin as a model

In 2002, Michael Grieves presented to industry the digital twin (DT) concept under the term of "Conceptual Ideal for PLM" at the University of Michigan for the formation of a Product Lifecycle Management (PLM) center (Grieves, 2016). More recently, the term "digital twin" was adopted by NASA (Piascik et al., 2010; Caruso et al. 2010; Tuegel et al., 2011; Glaessgen & Sturgel, 2012). A DT is "a set of virtual information constructs that fully describes a potential or actual physical manufactured product from the micro atomic level to the macro geometrical level" (Grieves, 2016). This definition usually refers to the structure of a product. It should be extended to its function. Figure 1.12 presents a conceptual map of the DT concept.

A DT is a virtual instance of a physical/cognitive system that enables to simulate dynamic phenomena of both structures and functions of a system. Madni, Madni, and Lucero (2019) claimed that **DT extends model-based system engineering.**[4] According to them, a DT can be used as a model of a real-world system to represent and simulate its structure, performance (i.e., function), health status, and mission-specific characteristics during the whole life cycle of the system and incrementally update it from experience (e.g., malfunctions experienced, maintenance, and repair history). In other words, a DT can be used as a recipient of experience

---

[4] INCOSE defines that "Model-based systems engineering (MBSE) is the formalized application of modeling to support system requirements, design, analysis, verification, and validation activities beginning in the conceptual design phase and continuing throughout development and later life cycle phases."

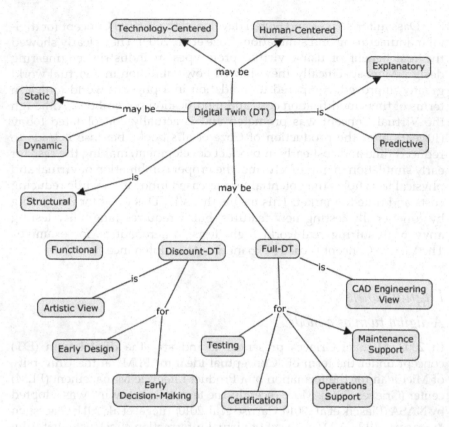

Figure 1.12 DT definitions and properties.

feedback information and support for system performance (e.g., preventive and timely maintenance based on knowledge of the system's maintenance history and observed system behavior). A DT is, therefore, a great support to improve understanding of the various relationships between system design and usages. In addition, a DT enables to support traceability and logistics along the whole life cycle of a system.

The EON system is a good example of early AI development of such experience feedback model in medicine (Musen et al., 1996). EON offers advice regarding the management of patients who are following clinical trial protocols for AIDS or HIV infection and breast cancer. EON is based on Friedland and Iwasaki's skeletal plans (Friedland & Iwasaki, 1985), an ontology framework, a computer-based patient record system, and it includes general-purpose software components that:

- Interpret abstract protocol specifications to make appropriate patient-specific treatment plans

- Infer from timestamped patient data higher-level, interval-based, abstract concepts
- Perform time-oriented queries on a time-oriented patient database
- Allow acquisition and maintenance of protocol knowledge in a manner that facilitates efficient processing both by humans and by computers.

Digital twin levels

Madni and his colleagues introduced four DT levels (Table 1.2).

Note that Pre-DT level 1 is the traditional virtual prototype created during upfront systems engineering. Madni et al. consider that a DT exists when a physical twin exists also; otherwise, it is a virtual prototype. Later on, a DT can include machine learning mechanisms (levels 3 and 4) that incrementally improve DT performance and support.

Predictive versus explanatory digital twins

Considering a DT as a system model, in the systems engineering sense, a distinction should be made between predictive DT and explanatory DT.

- A **predictive DT** is typically an analog to a very well-known model that works similarly to the DT. It is consequently short term, rigid, and focused on a specific process or phenomenon. It can be used for marketing and, in some specific cases, when we master the domain to operationally predict crucial system's states. A predictive DT is supported by a quantitative model.
- An **explanatory DT** is defined by specific ontology[5] (i.e., specific syntax and semantics). It is longer term, flexible, and domain specific. It can be used for analysis, design, and evaluation of complex systems. It can be used to document design and development process and its evolutionary solutions. An explanatory DT is supported by a qualitative model.

A DT is a digital model that enables running simulations to

- **Predict** behavior and performance of a real-world system
- **Explain** why the system behaves the way it behaves.

---

[5] At this point, an ontology can be represented by a taxonomic network of system concepts. We often talk about an ontology as a conceptual model.

Table 1.2 DT levels

Level	Model sophistication	Physical twin	Data acquisition from physical twin	ML (operator preferences)	ML (system/environment)
1. Pre-DT	Virtual system model with emphasis on technology/technical risk mitigation	Does not exist	Not applicable	No	No
2. DT	Virtual system model of the physical twin	Exists	Performance, health status, maintenance; batch updates	No	No
3. Adaptive DT	Virtual system model of the physical twin with adaptive user interface	Exists	Performance, health status, maintenance; real-time updates	Yes	No
4. Intelligent DT	Virtual system model of the physical twin with adaptive user interface and reinforcement learning	Exists	Performance, health status, maintenance, environment; both batch/real-time updates	Yes	Yes

Source: Madni, Madni, and Lucero (2019).

Predictive DT can be very effective to predict short-term issues or probable outcomes for a system. However, its validity must be always questioned since an analog has its limitations. Explanatory DT is a longer-term endeavor that human-centered designers of life-critical systems should always consider. It concerns design rationale and decisions history, evaluations of all kinds, and strategic changes with respect to tangibility testing (we will see later on how we can test tangibility).

Varieties of digital twin's purposes and components

A DT can be used during the whole life cycle of a system. It can be used for support of training, operations, maintenance, and eventually dismantling. Validation and certification of a DT is never finished. Experience feedback enables DT modification and evolution along the way by integrating new features, as well as modifying and/or removing old features. Considering system's life cycle, DTs can be used in a variety of industrial activities including:

- Product design
- Engineering optimization
- Smart manufacturing
- Job shop
- Scheduling
- Human–machine collaboration
- Operations diagnostic and decision-making
- Prognostics and health management
- Maintenance management
- More generally, product life cycle data management.

In any case, DTs require M&S, data management (sometimes big data), cybersecurity, interconnectivity, and interoperability (among several interconnected DTs). Key technologies for DT implementation are as follows (not an exhaustive list):

- System modeling and virtual prototyping support systems
- Simulation and HITLS
- Automatic control systems (cyber-physical systems)
- Internet of Things (IoT)
- Animations and video streaming systems
- Computer gaming engines (e.g., Unity 3D)
- AI and robotics
- Data science and machine learning
- Data visualization and human–computer interaction
- Tangible interactive systems (Boy, 2016).

Technology-centered versus human-centered digital twins

In addition, a DT can be seen from various viewpoints:

- Technology-centered
- Human-centered.

A **technology-centered DT** (TC-DT) is typically designed and developed to optimize internal functioning of the system it represents, and later on a user interface to enable it to be used. TC-DT results from the use of engineering approaches and methods – "this is a great algorithm, let's use it and we will see if people can use the resulting system later!" In other words, a TC-DT is an analog of an existing model. When well designed and developed, a TC-DT can be useful for prediction.

Instead, a **human-centered DT** (HC-DT) is designed and developed to be evolutionary (i.e., easily modifiable). It is designed to support HITLSs – like a video game – considering possible usages and observing the way it is used. Actual development of the whole system is eventually done later (the user interface being a component of the overall system). HC-DT development is typically agile. An HC-DT is a knowledge-based model of a system. It can, therefore, be useful for explaining observed behaviors and phenomena.

Therefore, the question is: a DT for doing what? You could develop a DT to test a very early concept within a **design team** that includes a group of experts with different backgrounds. Note that the design team includes targeted users or human operators of the system to be developed. Each member of the design team should understand the same thing as the others. This is the reason why they should have the same objectives and share the same SA of what is being designed and further developed. In this case, the DT is a concrete support to SSA. Each member of the design team can see the same thing and eventually manipulate it.

A DT for SSA starts with a **"discount DT."** This could be done, for example, with the help of an artist, capable of producing a DT in the form of a cartoon or animation of the targeted system to be developed. This is what was done during the design of NASA Space Exploration Vehicle (SEV)[6] during Blue Sky workshops carried out from 2006 to 2008. Discount DTs (DDTs) considerably increase SSA. A DDT is incrementally modified using critical thinking and experience feedback from DTMs. This explicit visualization is commonly used as a **mediating representation** useful and usable among the members of the design team.

---

[6] The SEV concept was initially called the Flexible Roving Expedition Device (FRED), then the Lunar Electric Rover (LER) and the Small Pressurized Rover (SPR).

Once the DDT is fully completed and approved by all members of the design team, it can be used to develop computer-aided design (CAD) of the system in depth. Using real numbers determining system's structure and function contributes to tangibilize the whole thing. The DT then becomes more rational and **tangible**. Once a "final" version is approved, physical construction of the system can start. This is another stage of tangibilization. Once a satisfactory version of the physical prototype is constructed, it can be tested in the "real world." In the SEV case, physical prototypes were tested in lunar-analog environments. Again, this is another stage of tangibilization. Each time the physical prototype is improved from experience feedback, its DT must be modified accordingly.

Human-In-The-Loop Simulation (HITLS): an aeronautical design case

HITLS is supported by a model. It enables people to interact within a DT simulating the system being developed. HITLS enables design teams to deeper study of people's activity inherent to the activity of the overall system being developed.

HITLS can also be called interactive simulations that are suited for analyzing large and semi-structured problems, especially in which human interaction is an important consideration. They "can be used for generating a better understanding of human behavior under complex situations by visually highlighting features that may not be readily accounted for in traditional simulations. Additionally, they can also be used for systems analysis under operational conditions as well as for simulator-based training" (Narayanan & Kidambi, 2011).

Let's illustrate the power of HITLS on an aeronautical design case. The Onboard Weather Situation Awareness[7] System (OWSAS), developed at Florida Institute of Technology,[8] is a good example of a system

---

[7] Situation awareness (SA) is commonly modeled as a pilot's ability to perceive, comprehend, and project a given situation (Endsley, 1995). Getting the right SA is difficult when we deal with life-critical systems in unpredictable situations, such as weather situations. Weather SA that aircrew can form can be supported by appropriate onboard 3D visualizations (Laurain, Boy & Stephane, 2015). We found that it is useful to merge several sources of data, including the National Climatic Data Center (NCDC), that hosts all National Oceanic and Atmospheric Administration (NOAA) meteorological data (Figure 1.13) and data from the National Weather Service (NWS). Other sources of weather data are available in Boy (2016).

[8] OWSAS was developed in my HCDi lab at FIT circa 2013–2017 and ended up supporting Sébastien Boulnois's PhD dissertation topic (Boulnois, 2018). Boulnois contributed to determine how weather information, presented in 2D and/or 3D, and associated with interaction features, could impact airline pilots' weather-related situation awareness and decision-making capabilities. He used HCD principles, including several knowledge elicitation, participatory design, Human-In-The-Loop Simulations, and evaluation sessions.

of systems,[9] since it involves data fusion of several weather data coming from various sources, visualization of aggregated data (e.g., 3D weather display), and integration into an aircraft cockpit environment. OWSAS[10] was designed and developed as a virtual prototype, and led to Sébastien Boulnois's PhD dissertation (Figure 1.13).

At each step of this iterative process, OWSAS has been proven to improve SA and it is expected this to be the case throughout its life cycle. The next iteration always involves decision-making, either to stop or to modify OWSAS for the test of an improved version. Improvements are not only software-based, but they can also be hardware-based (e.g., we needed to choose a different tablet after the first iteration). Maturity was also tested (e.g., a new kind of authority sharing between flight deck and ground was discovered after the second iteration). Of course, OWSAS was modified to consider these activity-based discoveries.[11] We typically talk about emergent properties. Based on several scenarios tested with HITLSs, we also

---

[9] Weather certainly provides one of the most important extrinsic uncertainties in aviation. It is very difficult to predict weather correctly with current technological means. Commercial aircraft are equipped with weather radar (located in the nose of the aircraft). This is good, but short term, information for pilots (i.e., you need to be close to the weather to get useful information). Weather is a natural system that causes 70% of all flight delays in the US and 23% of airplane accidents. Therefore, not anticipating weather conditions correctly leads to disastrous consequences for human lives, the economy (e.g., airlines lost around two billions of dollars per year due to weather issues (Weber et al., 2006)), and the environment (e.g., delays lead to increasing fuel consumption and pollution). This is considered in NextGen. In addition, this unpredictability of weather needs to be combined with increasing saturation of air traffic (i.e., this is a system-of-systems issues).

[10] OWSAS was first developed on a digital tablet, usable as a mobile device or a fixed instrument in the cockpit. Several versions of OWSAS were incrementally developed and tested with professional pilots on FIT's HCDi Boeing-737 and Airbus-320 flight deck simulators, using an agile approach.

[11] For example, main parameters were identified as being the angle of the layers, shapes of the 3D geometric forms (i.e., cylinders), and their colors and opacity. Verticality on planet Earth is always very small compared to horizontal distances, and the various layer angles were exaggerated by a factor of 3. Geographers commonly use two different factors depending on the maps: a factor of 1.5 for a mountain relief map (e.g., San Francisco) and a factor of 7 for deep water relief variations (Rumsey & Williams, 2002). Coastal environment experts prefer a factor of 3 (Milson & Alibrandi, 2008). Other authors directly integrate a cursor to modify these exaggeration factors depending on specific needs (Kienberger & Tiede, 2008). In order to simplify prototypes and pilots' interpretation, storm shapes were represented as cylinders, which are called Critical Storm Constraints. We used bright colors for the cylinders, consistent with NEXRAD data colors (e.g., red and orange). This two-color coding enables one to easily distinguish between two zones: dangerous zones (i.e., red) – where the storm is currently striking, and unadvised zones (i.e., orange) – where it could be dangerous to go in the near future. Cylinder opacity depends on the color, because the red cylinders needed to be easily detected but not too much; otherwise, they would hide the relief. Opacity of the data layers was also changed, allowing the relief to be observable across the first three layers (common altitude in flight). This transparency enables the pilots to see radar data (i.e., enough opacity to be seen and understood without effort) and orient themselves by looking at the relief, cities, and waypoints. Finally, the Flight Management System (FMS) flight path was added to the 3D model.

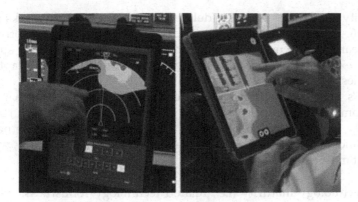

Figure 1.13 An experimental comparison of OWSAS 2D and 3D displays in a HITLS environment developed at FIT's HCDi Lab. (Photos provided by courtesy of Sébastien Boulnois.)

confirmed incrementally that OWSAS was a strategic system, in contrast with current tactical information provided by the aircraft nose radar.

Agile development and formative evaluations

Somebody is agile when he or she is able to create a solution that responds to change. Agility enables to deal with uncertainty and highly dynamic environments. An important corollary of agile innovative development is adaptation to environment. The Agile Manifesto (2015) is aligned with this definition of "Agile," and has four important goals:

1. Focus more on individuals and interactions than engineering tools and processes that are typically rigid (i.e., when a critical situation occurs, people involved should be ready to handle situations in a flexible way)
2. Working software is more important than comprehensive documentation
3. Customer collaboration is more vital than contract negotiation
4. The process should respond to change rather than follow a plan.

Agile development of software can be generalized to increasingly tangible cyber-physical systems[12] development. Main properties of agile development are the following: continuous delivery, customer satisfaction,

---

[12] Starting here, I will use the term "cyber-physical system" or CPS to denote a global concept (i.e., not limited to automatic control or automation meaning). Increasingly-tangible means that CPS being developed are incrementally tangibilized physically and figuratively (see further explanation on tangibility in Chapter 2).

flexibility for changes, concrete delivery at any time, system works at any time, life-cycled participatory design, face-to-face complicity, motivation, trust, team spirit, environment of appreciation, empowerment, sustainability, excellence, design for simplicity, self-organization, adaptation, usefulness, and usability.

Let's make clear that **"continuous delivery"** means that agile development promotes delivery of a tangible system at a given stage of maturity that needs to be identified. This is the reason why we need to better understand what maturity means (see Chapter 8). For now, we will say that maturity is a matter of:

- Technology maturity (i.e., related technology robustness, stability, controllability, and observability)
- Maturity of practice (i.e., related to people intentionality and reactivity)
- Societal maturity (i.e., related to culture and organizations).

In addition, "continuous" does mean that we are only interested in continuity in technology evolution but also technological ruptures and revolutions. Indeed, unlike HFE specialists of the 1990s who used to emphasize "automation surprises" using the agile approach, we like to discover such surprises by observing activities at design and development time. The more we will **identify** such **emerging properties**, the more we will gain in maturity (i.e., technology, practice, and societal).

Agile development is scenario-based (i.e., stories feed in virtual prototypes and test cases). In other words, orthogonal scenarios are incrementally defined and further tested using virtual prototypes and test cases:

- **Declarative scenarios** that define structural configurations (also called infrastructures)
- **Procedural scenarios** that define functional chronologies (also called stories).

Choice of such scenarios can of course be very difficult. At this point, scenario definition is more an art than a technique; it is based on expertise and experience. Indeed, storytelling requires expertise and experience from people who are subject-matter experts. It is always difficult and often impossible to quantify such expertise and experience, but it is extremely useful to qualify them to guide design decision-making processes.

For example, design and development of autonomous vehicles should not only be an egotistical individual goal but also a collective endeavor. Procedural scenarios should then be developed in which underlying societal models should be considered, simulated, and tested in real-world testbed environments. Following up the work developed by Jean-Pierre Orfeuil and Yann Leriche (2019), their TRUST model could be considered as a support for simulations to better understand current **evolution of autonomous vehicles and their potential usages**. TRUST means Technologies, Rules, Usages, Systems and services, and Territories.

You now understand that M&S can help figuring out these kinds of issues. Of course, M&S is done in a virtual world, and we need to clearly understand this aspect. When Jules Vernes[13] wrote "Around the world in eighty days," he anticipated this possible future, but he probably never imagined that François Gabart[14] would sail alone around the globe within forty-two days. Even better, astronauts currently turn around the world in about ninety minutes in the International Space Station. He did not have the means to simulate it, but he projected himself into the future. This is what we can do today using M&S. Computing power now enables us to project ourselves in a number of possible futures and figure out emergent behaviors and properties that could influence engineering design. I strongly believe that not going in this direction would be a mistake. Again, "tangibilization" of these possible futures is a matter of creativity and rationalization.

At this point, it is important to define a set of principles, metrics, measurement models, and observable variables to assess likelihood of these possible futures. This is where formative evaluations enter into play. The next chapter will emphasize tangibility problem and potential solutions.

Summary

When HSI for engineering designers is based on TCE, it naturally leads to corrective HFE; otherwise, when based on HCD, it leads to agile development and incremental formative evaluations (Figure 1.14). HCD-generated HSI brings positive design flexibility, more system knowledge, and improved resource commitment processes. HCD is typically based on DCs (already defined and described above), DTs, and HITLSs that enable discovery of emerging properties and functions through activity analysis.

[13] Vernes, J. Around the World in Eighty Days. Translated by Michael Glencross, with an introduction by Brian Aldiss. Published by Penguin Classics in 2004. ISBN 978-0140449068.
[14] https://fr.wikipedia.org/wiki/Fran%C3%A7ois_Gabart.

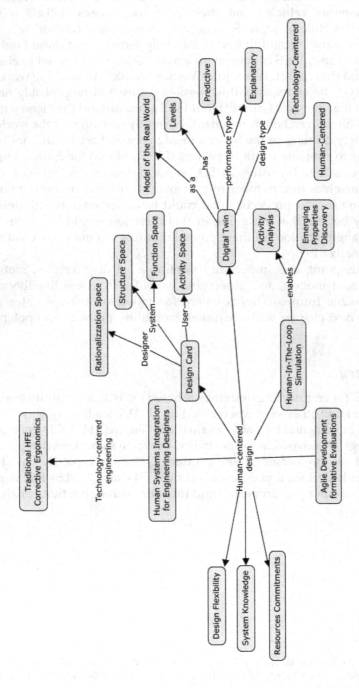

Figure 1.14 Engineering designers' view of HSI.

chapter two

Tangibility problems and potential solutions

> "Tangibility is the property that a phenomenon exhibits if it has and/or transports mass and/ or energy and/or momentum." A commonplace understanding of "tangibility" renders it as an attribute allowing something to be perceptible to the senses.
>
> **Piyush Mathur**
> *Philosopher (2017)*

Something is tangible when you can grasp it, hold it, feel it, understand it, and so on. Can we grasp virtual things in the way we grasp physical things? This question addresses what tangibility research is about. Industrial companies and everyday things are continuously becoming more digital. Some of us are getting lost in this growing virtual world, where socio-media invaded our lives to a point that requires more thoughts and investigations to clarify what are the causes of unexplained observed emerging behaviors.

Physical and figurative tangibility

For example, drone technology has recently developed for many kinds of purposes. Let's choose drones, which you can buy in consumer electronics stores, made for taking pictures and films, and also capable of delivering goods. Are they tangible? A first answer to this question could be to say that if such drones are able to evolve among people, other vehicles, and various kinds of infrastructures in a safe, efficient, and comfortable way for everyone, then they can be considered as tangible. They will be physically tangible if they can control themselves correctly in the physical environment around them. They will be figuratively tangible if they are useful and usable for many tasks that people would be able to assign to them (e.g., package delivery, image capture for news purposes, such as television, and disaster management support).

> Something is tangible when it is **graspable** in the **physical** sense but also **understandable** in the **figurative** sense (e.g., an idea or a concept that can be grasped by the mind).

Tangibility can be understood in many ways depending on cultures, organizations, and people personality, for example. Tangibility can be usefully associated with **safety, efficiency, and comfort**, which are culturally dependent. Is the tangibility concept universal? Obviously, operational tangibility is inherently dependent on governance (i.e., what is allowed to do and what is not). Indeed, some countries may have loose regulations compared to others. Somebody who lives in a country governed by a dictator will probably consider that hierarchical organizations provide models of figurative tangibility that are different from models understood by people who live in a democratic country. This is the reason why operational tangibility definition deserves more attention at the international level.

Inversion of engineering approaches from the 20th to 21st century

For a long time, technology-centered engineering was done **inside-out**, that is, designing and developing the technological kernel of a machine and finishing up by the user interface. In other words, inside-out engineering design goes **from means to purpose** (i.e., technology should be fully developed in order to be able to test possible activity, and then determine what kind of usages developed technology is really supporting). During the 20th century, it is not really surprising that we had to follow this kind of approach, because technology maturity, maturity of practice, and social maturity were not going as fast as what we are experiencing today using digital technology and data science.

Conversely, use of Virtual Human-Centered Design (VHCD) is done **outside-in**, that is, designing and developing virtual prototypes and digital twins, testing them in Human-In-The-Loop Simulations (HITLSs), using formative evaluation results to rapidly redesign in an agile way. Engineering design today can then go **from purpose to means** (i.e., we can test activity very early during the design process and incorporate testing results into designs). Figure 2.1 shows the paradigm shift from the 20th century, when automation was the main transformation (i.e., functionalization of structures), to the 21st century, which puts the tangibility issue to the front (i.e., development of tangible structures supporting purposeful functions).

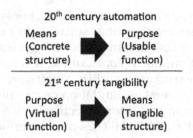

Figure 2.1 Paradigm shift: from "means-to-purpose" to "purpose-to-means."

It is crucial to realize that if automation was a 20th-century issue, tangibility is one of the major issues of the 21st century. Why? By definition, VHCD is based on virtual prototypes of systems being designed. It is now usual to talk about **digital twins**.

During the design process, if the use of digital twins clearly supports human-centered formative evaluations and, therefore, enables good HCD, it is done in the virtual world. Consequently, we need to verify that resulting designs are tangible from both figurative and physical viewpoints.

During the 20th century, engineering was focused on the creation of tangible assembly of tangible parts, themselves developed from tangible sub-parts and so on. In other words, engineering was technology-centered, mostly focusing on hardware means. Electronics and software were massively introduced late during the last period of the 20th century, to the point that we shifted from hardware to software. As an example, if you go to a garage because you have an issue with your car, the person in charge will come with a tablet, plug it to a USB socket under your steering wheel, and tell you how much the repair will cost. In most cases, he or she will not tell you what it is except if you ask him or her specifically. Hardware has become virtual and is processed as such for most of us. For example, cars were developed during the 20th century, sometimes almost independently from the development of road infrastructures. Car designs and road designs were not cross-fertilized, because both kinds of designs were engineering-centered and with Human Systems Integration (HSI). I would say even more; they were corporation-centered.

It is interesting to notice that since the beginning of the 21st century, almost any design is started on a computer screen, using PowerPoint, then modeling and simulating the system being designed toward the construction of a digital twin. We use software for doing this.

Incrementally, we tangibilize this virtual object or system to make a real physical thing. We are moving from software to hardware. We clearly see that VHCD is possible creating and using virtual prototypes; the main issue remains tangibility. For example, it is now possible to model cars and road infrastructures within a software environment composed of several interoperable simulation environments and play with the resulting simulation tool. More specifically, we can test driver–car cooperation (Bellet et al., 2010). HITLSs can be performed and assessed. Car designs and road designs can be cross-fertilized, in a human-centered way.

VHCD is scenario-based and supported by HITLS where human operators (e.g., drivers) are facing simulated situations that are as much tangible (i.e., realistic) as possible. Human operators are immersed into a realistic virtual environment, which in itself is a system of systems (e.g., car and roads structures and functions). Raw data are collected during HITLS sessions (e.g., eye tracking, control variable measurements) and further analyzed, typically by challenging them with other assessments (e.g., interviews, brainstorming sessions, subjective assessments). It is usual that assessments are a good mix of objective and subjective measures taken from a well-defined experimental protocol.

Tangibility attributes and metrics: the evaluation process

VHCD works on the basis of virtual prototypes and digital twins of systems being designed and further used. Virtual prototypes are incrementally developed in an agile way and assessed using metrics, measurement models, and observable variables. Digital twins are further used at operations time to help making diagnostics, decision-making, and action taking on systems being operated.

Metrics are high-level attributes that provide **meaning** to an assessment. For example, "human workload" is a metrics that provides meaning on human performance deficit (e.g., going from lack of vigilance to excess of stress) or human errors. Workload deals with resource management (e.g., excess workload goes with resource decrease). Measuring workload is a challenge and several approaches and models have been developed. Examples of such measurement models are the (required time/available time) ratio (an aeronautical example will be provided in Chapter 8) and Cooper–Harper subjective rating scale (e.g., between 1 for "no workload" and 10 for "unmanageable workload") (Cooper & Harper, 1969) (Figure 2.2). Each measurement model requires sensing or acquiring observable variables such as "required time," "available time," and subjective values provided by human operators.

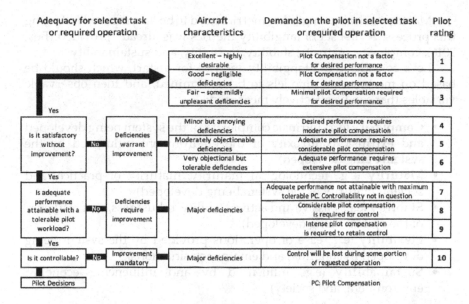

Adequacy for selected task or required operation	Aircraft characteristics	Demands on the pilot in selected task or required operation	Pilot rating
	Excellent – highly desirable	Pilot Compensation not a factor for desired performance	1
	Good – negligible deficiencies	Pilot Compensation not a factor for desired performance	2
	Fair – some mildly unpleasant deficiencies	Minimal pilot Compensation required for desired performance	3
Is it satisfactory without improvement? **No** Deficiencies warrant improvement	Minor but annoying deficiencies	Desired performance requires moderate pilot compensation	4
	Moderately objectionable deficiencies	Adequate performance requires considerable pilot compensation	5
	Very objectional but tolerable deficiencies	Adequate performance requires extensive pilot compensation	6
Is adequate performance attainable with a tolerable pilot workload? **No** Deficiencies require improvement	Major deficiencies	Adequate performance not attainable with maximum tolerable PC. Controllability not in question	7
		Considerable pilot compensation is required for control	8
		Intense pilot compensation is required to retain control	9
Is it controllable? **No** Improvement mandatory	Major deficiencies	Control will be lost during some portion of requested operation	10

Pilot Decisions   PC: Pilot Compensation

Figure 2.2 Cooper–Harper handling qualities rating scale (NASA TND 5153).

Cooper–Harper subjective rating scale is a good example of support for assessing figurative tangibility as subjectively provided by human operators controlling and/or managing complex systems. It has been extensively used in aeronautics in many commercial aircraft certification campaigns.

Once data are acquired from observable variables, interpretations are made to derive meaningful information and end up with knowledge that can be used to upgrade currently used metrics (Figure 2.3). This cycle is run until satisfactory metrics, measurement models, and observable variables are found. The ultimate goal is the derivation of appropriate standardization of this kind of evaluation process. Note that this evaluation method standardization process is valid for any type of metrics whether they are physical or cognitive.

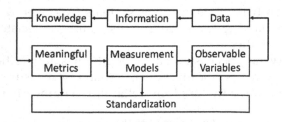

Figure 2.3 Evaluation method standardization process.

More specifically, tangibility metrics need to be decomposed following this process. A first set of tangibility metrics was already provided (Boy, 2016): complexity, maturity, stability, flexibility, and sustainability.

Let's consider these tangibility metrics for a start, which should be based on measurement models to be determined, and then observable variables that can be effectively measured:

- **Complexity** (e.g., intrinsic complexity of the system being developed and extrinsic complexity of the environment influenced by the system being developed)
- **Maturity** (e.g., technological maturity, maturity of practice, and societal maturity of the system being developed)
- **Stability** (e.g., passive and active stability, robustness, and resilience of the system being developed)
- **Flexibility** (e.g., ease of operations provided by the system being developed in abnormal and emergency situations)
- **Sustainability** (e.g., influenced by and influencing economy, environment, and society).

An assessment of tangibility may be required if you wish to determine whether you have an acceptable system to be delivered in the real world; if you have enough capacity for additional tasks and how much; or whether you need to provide additional resources to cope with emergencies, incidents, or process upsets. Tangibility should be assessed anytime a new system is designed and new roles and responsibilities are anticipated or discovered through HITLS.

From meaningful metrics to measurement models to observable variables

With massive development of **increasingly autonomous systems**[1] (both human and machine systems), meaningful metrics need to be developed. Let's try to describe what **trust**, as a meaningful metrics within a multi-agent system (or a system of systems), could developed in terms of specific measurement models and observable variables. Trust in automation was studied for a long time (Lee & See, 2004). More research has been started on trust in autonomous systems (Atkinson & Clark, 2013; Atkinson, Friedland & Lyons, 2012). Trust can be associated with reliability and tangibility.

Reliability has been studied in depth in safety-critical systems engineering. More specifically, system reliability measures have been developed and broadly used in reliability engineering (Laprie &

---

[1] US National Academies produced a report on that topic in the aviation sector (Clarke & Lauber, 2014).

Kanoun, 1996). Human reliability was also covered for several decades (Hollnagel, 1993). Since its inspection, human reliability assessment (HRA) was based on the human error assumption associated with individual actions. HRA was an extension of probabilistic safety analysis (PSA) that considered the human element. Unfortunately, HRA did not work properly because human and machine functions could not be considered and modeled in the same way. Context matters! People do not react always rationally context-wise. Most importantly, people tend to invent ad hoc appropriate solutions adapted to context, which machines cannot do. This is the reason why HRA research did not continue to be developed.

Let's consider now figurative tangibility as a close concept to the concept of **trust** (e.g., "I trust you because what you say makes sense, what you say is tangible, or has a meaningful and graspable reality"). Consequently, trust metrics could be developed by considering the following tangibility properties: complexity, maturity, stability, flexibility, and sustainability. Table 2.1 provides an example of decomposed trust metrics in the context of human–machine teaming (i.e., an environment where humans and machines cooperate toward a common goal).

Table 2.1 Trust metrics using tangibility attributes in the context of human–machine teaming

Trust meaningful metrics	Measurement models	Observable variables/methods
Complexity	Familiarity with multi-agent interaction (teaming competence) Delegation competence (cognitive function transfer)	From cognitive function analysis, use functions and links as complexity factors Assess function allocation among human and machine agents in terms of competence in context Assess complexity coming from emergence of new cognitive functions after cognitive function transfer (automation or autonomization)
Maturity	Technological, organizational, and human operator maturity (TOP model) Stabilization of emergent functions	Usability and usefulness assessment (usability engineering methods) Resilience engineering assessment (FRAM, Hollnagel, 2014)

(Continued)

Table 2.1 (Continued) Trust metrics using tangibility attributes in the context of human–machine teaming

Trust meaningful metrics	Measurement models	Observable variables/methods
Functional stability (avoiding surprises in the context-resource hyperspace)	Context of validity Availability of required resources Outcome satisfaction	Use results of cognitive function analysis to find out repercussion of getting out contexts of validity in the use of functions Same to find out repercussion of lack of resources (e.g., in cases of malfunctions)
Flexibility	Mastery of procedure following and automation (understanding contexts) Problem identification and solving (cognitive support)	Study effectivity and efficiency of tools and learning requirements in case of getting out of validity contexts Easy of problem stating in abnormal and emergency situations
Sustainability	Reliability in time "I trust you as long as you keep your promises!"	Human and machine fatigue Human/machine health monitoring Change management

There is another way of decomposing trust metrics (adapted from Steinfeld et al., 2006), considering the overall human–machine system performance, human operator performance, and increasingly autonomous machine performance (Table 2.2).

Many other metrics, measurement models, and observable variables can be developed with respect to the problem to be solved.

Incremental construction of tangible things: an ontology-based method

Following up the current trend of designing and developing systems that is going from software to hardware, HSI adopts the outside-in design approach. This approach leads to designing first the user interface that includes the simulation of the overall system. Everything is, therefore, about software! Even hardware parts will be simulated by software. But before developing virtual prototypes, an **ontology** of the domain should be developed to get a common framework and language that will support

Table 2.2 Trust metrics of human–machine teaming performance

Trust meaningful metrics	Measurement models	Observable variables/methods
Human–machine system performance	Quantitative performance • Effectiveness • Effectivity	% mission accomplished with the designed autonomy Number and duration of human operator's interventions Time required to complete a task Number of tasks completed within a time period
	Subjective ratings Appropriate utilization of mixed initiative • Percentage of requests from automation assistance • Percentage of request from assistance made by operator • Number of interruptions of operator rated noncritical	Cooper–Harper rating scale use Cognitive function analysis For each function, assess human factors attached to it (memory load, visibility load, audibility load, etc.) Human-automation effort Distraction due to unnecessary incomprehension (why does it behave that way?)
Human operator performance	1. Situation awareness 2. Human performance and workload 3. Accuracy of mental models of device utilization	SAGAT (Endsley, 1995) NASA TLX (Hart & Staveland, 1988) Cardiac and respiratory measures (Wilson, 1992, 2001) Affordance identification Human operator's expectations Stimulus response compatibility Human–computer interaction (HCI) and engineering design measures (Newman & Lamming, 1995; Sanders & McCormick, 1993)
Increasingly autonomous machine performance	1. Self-awareness 2. Human awareness 3. Autonomy	Understanding intrinsic limitations Capacity for self-monitoring Self-recognition deviation from nominal Effectiveness at detecting, isolating, and recovering from faults Recognition of human behavior and deviation from nominal Capacity to appropriately correct human errors Neglect tolerance to measure autonomy (Goodrich & Olsen, 2003)

the design team and all stakeholders during the whole life cycle of the system being developed. This is the reason why HSI of complex systems, and more specifically increasingly autonomous complex systems, should be supported by artificial intelligence (AI) methods and tools.

In AI, an ontology is an axiomatic specification of concepts, characterizing a domain, by allocating terms (symbols) to them and relationships among them. There are several software tools that can be used to develop an ontology. The most commonly used tool to develop and maintain ontologies is *Protégé* (Musen, 2015; Horridge et al., 2011). More specifically, the web-based version of Protégé is very easy to use in a collaborative authoring manner. Other applications are available, such as CmapTools, to model and share knowledge (Novak & Cañas, 2006). Concept maps (Cmaps) are graphical representations of knowledge. CmapTools enable organizing these representations, which include concepts and relationships between concepts indicated by a connecting line linking two concepts.

Why should we develop an ontology? According to Noy and McGuinness (2019), reasons are usually the followings:

- Sharing common understanding of information structure among people or software agents
- Enabling reuse of domain knowledge
- Making domain assumptions explicit
- Separating domain knowledge from the operational knowledge
- Analyzing domain knowledge.

Developing an ontology is performing an exploration and an analysis of the domain we want to study. Basically, we want to improve our understanding of the various domain-entity types (e.g., executant, team, supervisor, mediator, coordinator), types of relationships among these entities (e.g., delegation, cooperation, authority, processing, temporal, spatial), and the various situations that make sense to be considered. We could say that an ontology is a conceptual model of the domain it represents (i.e., the way we see the domain) that should be tractable by a computer program. Of course, such a conceptual model should be elaborated by a team of domain experts and incrementally tested against the real world. It is clear that each team member has a specific perspective for elaborating and testing the conceptual model. For example, when an artist paints a landscape, he or she has his or her own perspective. Consequently, when we develop an ontology, a maximum of perspectives should be integrated within the conceptual model (i.e., the ontology). Experts usually are very good at telling stories (see Chapter 8). This is the reason why it is crucial to let them tell these stories; guide their generation; and capture meaningful domain entities, relationships among them, and related situations. This is where physical and cognitive function analysis takes place, and invariants are detected (incrementally by induction).

As an example, we are developing a model and simulation of a fleet of robots on an offshore oil-and-gas platform (Rolos, Masson & Boy, 2019). We decided to start designing an operations room that would enable people to operate the fleet of robots. We first elicited engineering and operational concepts from subject-matter experts. We formalized existing control rooms' design and use (oil-and-gas domain ontology), and planned on extending them to the management of a fleet of robots (problem-specific ontology). We incrementally developed a structural and functional ontology to identify and allocate current and emergent functions. This approach leads to the development of a software simulator that includes structures and functions of a virtual oil-and-gas platform, a virtual fleet of robots, and the related operations room. During the simulation, the various human and machine agents (represented as systems) are observed using specific measurement protocols and tools. An activity analysis is carried out and recommendations generated. These recommendations include various kinds of changes, including model changes, and possibility of replacing a software part (e.g., a robot or platform part) by its hardware equivalent on a physical site. We then incrementally tangibilize designed objects and agents. Tangibilization does not mean only substituting a virtual object or agent by its physical equivalent but also possibly modifying the model of it and, therefore, its physical design (Figure 2.4).

Example of an ontology of a robot

What is a robot? Most people would say a humanoid, like in Star Trek. However, there are many kinds of robots such as drones, autonomous cars, and so on. They are all made of hardware (supporting its structure) and software (supporting its functions). A robot is a system defined by its cognitive and physical structure of structures and function of functions.

A robot can also be defined as an agent (we already discussed the equivalence between system and agent). In turn, agents can be human, animal (e.g., a bird), or artificial (e.g., a robot). An agent can perceive, decide, and act on its environment. It has sensors, an information processor, and effectors. It can be:

- Goal-driven (i.e., intentional, take initiative, and recognize opportunities)
- Event-driven (i.e., reactive and able to keep connectivity with the environment and respond to changes)
- Socially driven (i.e., interactive and cooperative with other agents)
- Rational (i.e., programmable and making decisions)
- Adaptable (i.e., able to learn).

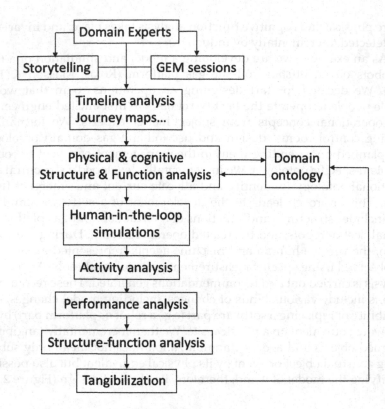

Figure 2.4 Ontology-based VHCD method.

A robot has a structure of structures and a function of functions. A non-exhaustive conceptual model of a robot is presented in Figure 2.5.

In addition to this declarative ontology (e.g., robot ontology presented on Figure 2.5), we need to define a set of scenarios, which can be qualified as procedural ontology. These scenarios will provide tasks that should be performed by the various agents involved. We will see in more detail how we can transcribe these scenarios using timelines and journey maps in Chapter 9. At this point, an example of scenario for the robot is provided in Table 2.3 (e.g., the human operator needs the position of a valve and requests the robot to get this information – the robot is equipped with a camera). In this very simple example, several assumptions are made, such as, the camera works fine, the Cloud information system works fine, and the robot is capable of rolling and climbing stairs.

Management of a fleet of robots on an oil-and-gas platform requires the specification of a network of robotics concepts. Here again we need to elicit concepts from both experienced people in the oil-and-gas domain and experts in robot remote control and management.

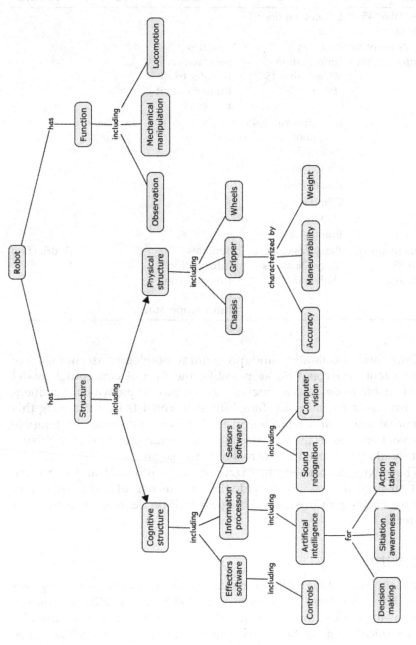

Figure 2.5 A non-exhaustive ontology (conceptual model) of a robot.

Table 2.3 Example of a timeline for robotic remote control and management

Human operator	Robot	Cloud	Valve-45	Camera
Needs Valve-45 Position	Located on floor-1			
Requests robot to get information	Receives information: "Get Valve-45 Position"	Mediates information transfer from human operator to robot		
	Identifies Valve-45 locations (Floor-2)		Located on Floor-2	
	Go to Floor-2			
	Roll to stairs			
	Climb stairs to Floor-2			
	Roll to Valve-45			
Gets an image of Valve-45 Position	Activates camera and visualizes Valve-45	Mediates information transfer from camera to human operator		Works fine

Once both declarative and procedural ontologies (scenarios) are as consistent and complete as possible, the overall conceptual model should enable to develop a concrete virtual prototype, which incrementally becomes a digital twin. Tangibilization consists in fine tuning this conceptual model and replacing virtual objects and agents by tangible equivalent entities in the real world. In this example above, the virtual robot should be replaced by the real one, and so on.

The next step consists in carrying out a physical and cognitive function analysis that will provide the various useful and usable functions for each agent involved. This part will be developed in the next chapter.

Summary

Early 21st-century tangibility problems are about sociotechnical digitalization that comes from a paradigm shift that inverts traditional concept of means-to-purpose to a new concept of purpose-to-means (Figure 2.6). Sociotechnical digitalization makes emerge tangibility issues that can be physical and/or figurative, and requires tangibility models and metrics

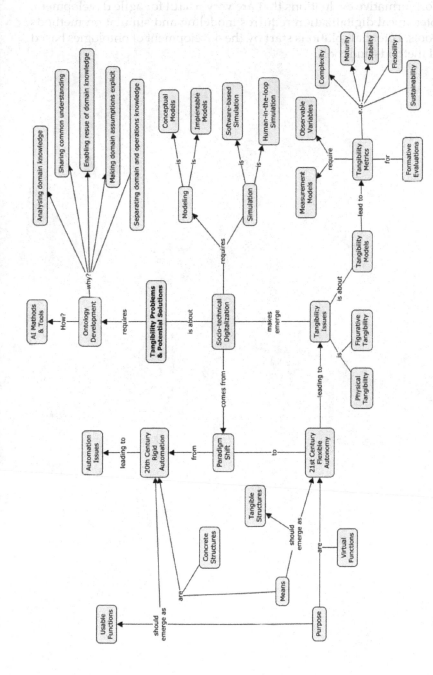

Figure 2.6 Tangibility problems and potential solutions.

(e.g., complexity, maturity, stability, flexibility, and sustainability), which support formative evaluations that are very useful for agile development. Sociotechnical digitalization requires modeling and simulation methods and tools. Potential solutions start by the development of ontologies based on AI methods and tools.

chapter three

Technology, organizations, and people

> Talent wins games, but teamwork and intelligence win championships.
>
> **Michael Jordan**
> *American former Professional Basketball Player.*

People and machines are systemic parts of a global system. Technology is supposed to be developed to improve safety, efficiency, and comfort, but humanity is barely considered when the goal is only financial. Financial optimization forces the development of systems that process and manage more things, faster and cheaper. Personnel constantly decreases in companies (e.g., consider the effect of car manufacturing robotics). Even if business is usually better financially, people's lives are not necessarily better in terms of employment and involvement in the development process. This is the reason why technology, organizations, and people have to be seriously considered during the whole life cycle of a system. More specifically, technology should be considered as a support to people and not the other way around. The question of authority needs to be addressed when designing and developing a complex system where people and machines have to collaborate. Who is in charge? What cultural shift results from the development and use of a new technology?

When technology is a matter of automation, some employees may be frightened because they will be shortly replaced by a machine. In other words, they will lose their jobs. Consequently, organizational scenarios and Human-In-The-Loop Simulation (HITLS) should be investigated before it is too late to find out what kinds of new jobs should emerge from the use of new technology. **Technological evolution is inevitable**. Therefore, employees should be involved in the transformation and not informed at the last minute that they will be fired! People should be aware of this evolution and participate in the change management process. We will review what are the relationships between technology, organizations, and people when we try to design and develop new ways of doing things.

Figure 3.1 The TOP model.

The TOP model

The TOP (technology, organizations, and people) model was developed during early 2000 to provide design teams with a framework that considers these three entities from the beginning of design and all along the life cycle of a system (Figure 3.1).

This is crucial when we are designing life-critical systems (Boy, 2013a). Let's take the case of aircraft automation evolution.

Technology will gain a lot in usefulness and usability by being developed in a participatory way (i.e., involving everybody within the organization). Usefulness and usability are improved when use of technology is easy to learn and stabilize, when human errors are easy to overcome, and when technology provides a great user experience in terms of satisfaction and effectiveness. Technology acceptability depends on such usefulness and usability, as well as social habits and culture that are sometimes difficult to change.

Organization should be investigated from two viewpoints:

- Structure (i.e., topology of the organization)
- Function (i.e., workflow generated within the organization).

An organization is a system of systems. Its structure can be hierarchical or heterarchical, for example. In human-centered design (HCD), I already explained the shift from the traditional army model, with a general at the top and soldiers at the bottom, to the **Orchestra model** (Boy, 2013a), with musicians, some of them being conductors and compositors. More formally, playing a symphony, the orchestra organization requires five kinds of components:

1. Music theory that is the common language
2. Scores produced and coordinated by composers
3. Workflow coordinated by a conductor
4. Musicians performing the actual symphony
5. Audience listening the produced symphony.

In this book, we will adopt the Orchestra model as a metaphor for any kind of contemporary systemic organization. Organizational issues will be analyzed in terms of communication, cooperation, and coordination, considering two orthogonal system attributes: structure and function.

People are the various stakeholders of organizations around the technology that is considered. They can be designers, engineers, developers, certifiers, maintainers, operators or users, trainers, and dismantlers (not an exhaustive list). People, in the TOP model, have activities and jobs. Anytime technology and/or organization change, people may change their activities and/or jobs. Sometimes, new technology may lead to people losing their jobs or conversely new jobs (i.e., functions) should be created, and therefore, a new set of people might be hired (i.e., a new structure should be created within the organization). People have their own human factors issues, such as fatigue, workload, physical and cognitive limitations, and creativity.

Human–Systems Integration (HSI) requires that we define the job of human systems integrators. This new job can be twofold whether it is defined before or after delivery of a system. Before delivery, the human systems integrator is a composer (using the Orchestra metaphor); after delivery, the human systems integrator is a conductor. In both cases, the human systems integrator is typically a member of the senior engineering staff reporting to either the systems engineering lead or chief engineer.

The TOP model is a simple framework that supports **co-design** of TOP's activities and jobs. For short, the TOP model frames HCD to insure better HSI. However, we need to go into more detail and consider other entities. More specifically, two entities, task and situation, should be considered in addition to the TOP model. The following section proposes the AUTOS pyramid concept, encapsulated within the TOP model, as a more sophisticated framework and method supporting design team activities.

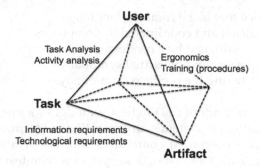

Figure 3.2 The AUT triangle.

The AUTOS Pyramid

The AUTOS pyramid concept was developed as a pragmatic and operational guide for HCD (Boy, 2011). "A" means Artifact (i.e., Technology in the TOP model), "U" means User (i.e., People in the TOP model), "T" means Task, "O" means Organization, and "S" means Situation.

The AUT triangle (Figure 3.2) associates:

- Artifacts may be cars or consumer electronic systems, devices, and parts, for example.
- Users[1] may be novices, experienced personnel, or experts, coming from and evolving in various cultures. They may be tired, stressed, making errors, old or young, as well as in very good shape and mood.
- Tasks may vary from high to low level (e.g., managing a team or an organization, designing, making decisions, and handling quality control). Each task involves one or several cognitive and/or physical functions that related users must learn and use.

Technical documentation complexity is very interesting to be tested because it is directly linked to the explanation of artifact complexity. The harder a system is to use, the more related technical documentation or performance support is required in order to provide appropriate assistance at the right time in the right format.

The AUT triangle enables the explanation of three edges:

- Task and activity analysis (U-T)
- Information requirements and technological requirements and limitations (T-A)
- Ergonomics and training (procedures) (T-U).

---

[1] Even if I prefer to use the term "people", tradition in human-machine systems is to use "human operators," especially in the process control and human engineering community, or "users," in the HCI community.

AUT complexity is characterized by content management, information density, and ergonomics rules. Content management is, in particular, linked to information relevance, alarm management, and display content management. Information density is linked to decluttering, information modality, diversity, and information-limited attractors, that is, objects on the instrument or display that are poorly informative for the execution of the task but nevertheless attract user's attention. The "PC screen do-it all syndrome" is a good indicator of information density (elicited improvement factors were screen size and zooming). **Redundancy** is always a good rule whether it repeats information for crosschecking, confirmation or comfort, or by explaining the "how," "where," and "when" an action can or should be performed. Ergonomics rules formalize user friendliness, that is, consistency, customization, human reliability, affordances, feedback, visibility, and appropriateness of the cognitive functions involved. Human reliability involves human error tolerance (therefore, the need for recovery means) and human error resistance (therefore, the existence of risk to resist to). In contrast with these negative side of people, we also need to investigate human engagement and involvement. To summarize, A-factors deal with the level of necessary *interface simplicity, explanation, redundancy,* and *situation awareness* that the artifact is required to offer to users.

The AUT triangle is limited to the local articulation of artifact, user, and task. It should be put in perspective within a targeted organizational environment, which includes all team players who/that will be represented as systems (in the next chapter) and also called "agents," whether they may be humans or machines, interacting with the user who performs the task using the artifact. Introduction of the organizational environment contributes to consider three additional edges that shape the AUTO tetrahedron (Figure 3.3) associating

- Social issues (U-O)
- Role and job analyses (T-O)
- Emergence and evolution (A-O).

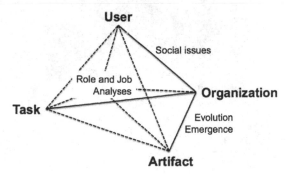

Figure 3.3 The AUTO tetrahedron.

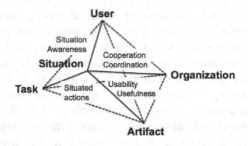

Figure 3.4 The AUTOS pyramid.

The AUTOS pyramid framework (Figure 3.4) is an extension of the AUTO tetrahedron that introduces a new dimension, the "Situation." The three new edges are

- Usability/usefulness (A-S)
- Situation awareness (U-S)
- Situated actions (T-S)
- Cooperation/coordination (O-S).

Clarifying the concept of situation

A situation can be viewed in many ways. A situation may refer to a dynamic set of states including multiple derivatives, in the mathematical sense. Let's try to construct a model of the various kinds of situations (Figure 3.5).

Ideally, the real world is characterized by an infinite number of highly interconnected states. This is what we call the "real situation." It may

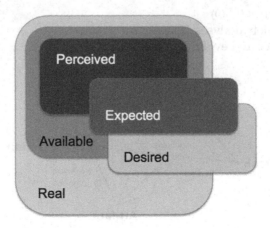

Figure 3.5 Various kinds of situations.

happen that some of these states are not available to us. For example, many states describing aircraft engine health are not directly available to pilots. States available to a human observer define the "available situation" (e.g., aircraft engine health states available to pilots). Note that the "available situation" is typically part of the "real situation." In addition, the "available situation" may not be totally perceived by the observer. What he/she perceives is called the "perceived situation." Of course, the "perceived situation" is part of the "available situation" but is also directed by what is being expected. The "desired" situation typically expresses a goal-driven behavior (e.g., we want to get to this point). The "expected" situation expresses an event-driven behavior (i.e., we anticipate a set of states to happen).

When people expect something to happen very strongly, they may be confused and mix the "perceived situation" with the "expected situation" (i.e., this is usually related to cultural context, distraction and focus of attention). There is huge difference between monitoring activities and control activities. People involved in a control activity are goal-driven. Their situation awareness process is directed by the task they need to perform (i.e., the role they have in the context where they are). Conversely, people who only have to monitor a process (and who do not have to act on it) need to use, and sometimes construct in real time, an artificial monitoring process that may be difficult, boring, and sometimes meaningless. In this second case, the situation awareness process has many chances not to be accomplished correctly.

Finally, the "perceived situation" is not necessarily a vector of some available states but a model or image that emerges from a specific combination of these states, incrementally modified over time. This is called experience acquisition. Human operators build their own mental models or mental images of the real situation. This mental image depends on people, cultural context, current people's activity, and other factors that are specific to the domain being studied. You can see here the influence of cognitive context on physical context, because what people perceive is not entirely the real situation but something constructed from the available situation and their own desired situation and background knowledge and skills.

Consequently, people, who are not familiar with complex situations in laboratory setups, may produce false interpretations one day or another. For this reason, HCD formative evaluations dealing with complex systems require training, minimal experience acquisition, and longer involvement of human operator subjects. This is the reason why I always recommend design teams to develop HCD processes based on real-world experimental setups (e.g., realistic aircraft simulators and professional pilots).

This leads to the definition of two more types of situation: intrinsic and extrinsic. Intrinsic situations are related to the complexity of human operators' capabilities. Extrinsic situations are related to the complexity of human operator's environment. Both types of situation could be expressed in terms of number of states and interconnections among these states. In both cases, appropriate models need to be developed. Real and available situations are categorized under the concept of extrinsic situations. Expected and desired situations characterize the concept of intrinsic situation. Perceived situations belong to both concepts of extrinsic and intrinsic situations.

Interdependency of AUTOS factors

We should realize that when we design and develop a system (i.e., an artifact), we also design and develop new jobs (i.e., user's profiles). Both artifact and user require a workable framework defined by tasks, as well as organizational setups and a large set of situations that define the environment. By using the AUTOS pyramid, a design team is equipped with a framework that enables guiding HCD, insuring completeness of methods, forcing robustness and resilience, and so on. Since this book is devoted to support designers and engineers in the design and development of complex human–machine systems, machine factors will not be developed from an engineering viewpoint, but a usage viewpoint.

The AUTOS pyramid has to be seen in the context of our growing information society, where Industry 4.0 starts to dictate our digital life. In the introduction, the evolution of engineering-oriented human-centered fields of investigation was described (Figure 0.2): from human factors and ergonomics (HFE) to human–computer interaction (HCI) to Virtual HCD leading to HSI.

Office automation, which started with text processing and the desktop metaphor, shaped HCI since the early 1980s, to the point that usability often refers to the ability to use a graphical interface that includes menus, buttons, windows, and so on. In contrast, process control automation was driven by control theories where feedback is the dominant concept and took care of real-time continuous processes such as nuclear, aerospace, or medical systems where time and dynamics are crucial issues together with safety-critical issues; human operators are skilled and expert professionals because industrial processes they control are life-critical and complex. Conversely, HCI developed the interaction comfort side; end users are everybody. HCI specialists got interested into learnability, efficiency, easy access to data, direct manipulation of metaphoric software objects (widgets), and pleasurable user experience, for example.

Talking about familiarity, people have become familiar with computer screen menus, windows, and interaction devices. HCI tremendously contributed to improve familiarity with computing systems. Furthermore, it is interesting to see how people's familiarity with successful interaction styles (e.g., interaction on smartphones) is expected on other systems. Indeed, most systems are now digitized and interactive.

When I was working on the premises of the Airbus 380 cockpit, I was incidentally told that it was going to be an "interactive cockpit." I thought about it and said that aircraft have been interactive for a long time. Indeed, pilots manipulated throttles and yokes to fly aircraft... this was interactive! However, someone told me that now cockpits were interactive because they have a pointing device and a keyboard! As a matter of fact, pilots do not interact with the mechanical pieces of their aircraft but through interactive computers. This (r)evolution started with the glass cockpits in the mid-eighties; we were talking about "fly-by-wire." More recently, the car industry introduced "drive-by-wire." More software was introduced incrementally, interconnecting computing systems among each other.

Of course, there are many other ways to develop and use a useful framework for HCD. We choose the five AUTOS dimensions because they have been proved to be very useful to drive HCD and categorize HSI complexity into relevant and appropriate issues. These aspects include design methods, techniques, and tools.

Articulation of structures and functions

Human–machine interaction (HMI) could be presented by describing human factors, machine factors, and interaction factors. We will see in Chapter 4 that a system can be defined as

- A system of systems
- A representation of a human or a machine
- A structure of structures associated to a function of functions.

Therefore, HSI does not only consider systems interacting among each other (i.e., what we usually call HMI, human–human interaction (HHI), or machine–machine interaction (MMI) but also the **topology** of a system of systems that can be transformed into an architecture (i.e., the structure of the system of systems). This systemic topology does not depend only on static decomposition of a system into sub-systems but also on dynamic function allocation onto system's infrastructure.

From an HSI point of view, system factors include structure factors and function factors. Users and artifact can be associated to structures. Task, organization, and situation can be associated to functions. System design is about articulating structures and functions to make a coherent whole.

System complexity can be split into intrinsic complexity and extrinsic complexity. Intrinsic complexity is complexity of a system of systems. Extrinsic complexity is complexity of the environment of a system (i.e., including organizational and situational environment in the AUTOS pyramid sense). The following concepts characterize system complexity (Figure 3.6):

- **Flexibility** that can be decomposed into intrinsic flexibility (i.e., ease of change within structure and function of systems within the system of systems) and extrinsic flexibility (i.e., ease of change of

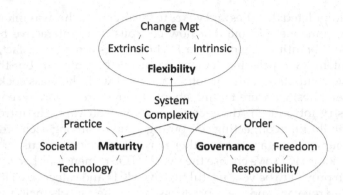

Figure 3.6 System complexity in terms of flexibility, maturity, and governance.

the overall structure and function of the system of systems within its organizational and situational environment).

- **Maturity** that can be understood by the incremental discovery of emergent features (e.g., articulations and integrations of structures and functions within the system of systems) – instead of piling up new functions and/or structures, new integrated structures/ functions are generated replacing the old ones – technology maturity is developed in parallel with maturity of practice (people's experience) and societal maturity (culture evolution).
- **Governance** that involves various types of systemic interaction models, which will be presented in Chapter 4, authority sharing models going from hierarchical to heterarchical (e.g., from the Old Army model to the Orchestra model), and responsibility allocation among stakeholders – governance usually stabilizes between free-dom (i.e., agent's autonomy) and order (i.e., coordination of autono-mous agents).

Summary

The TOP model is about technology (i.e., life cycle, usability, and useful-ness), organization (i.e., structures and functions, type of organization), and people (i.e., involved in participatory design and are designers, HSI specialists, engineers, human operators, certifiers, trainers, maintainers, and dismantler – not an exhaustive list). The TOP model can be detailed as the AUTOS pyramid framework that provides a large set of methods and tools for the analysis, design, and evaluation of systems being developed. The TOP model leads to studying system complexity in terms of flexibil-ity, maturity, and governance (Figure 3.7).

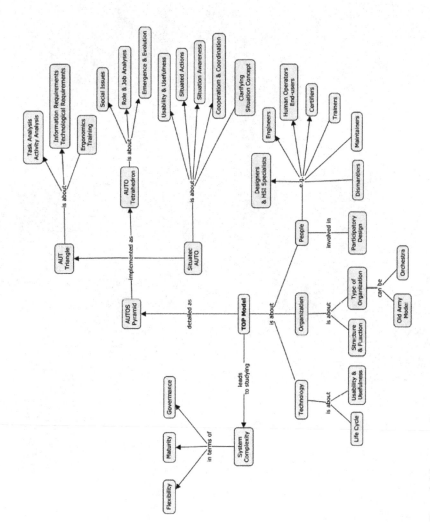

Figure 3.7 The TOP model framework.

chapter four

Formalizing Human–Systems Integration

> Human capabilities, skills, and needs must be considered early in the design and development process, and must be continuously considered throughout the development lifecycle.
>
> **David J. Fitts**
> *Space Architect*

Human–Systems Integration (HSI)[1] is both a process and a solution during the whole life cycle of a system, from its inception to its dismantling. As a process, HSI is a set of principles, methods, and tools that enable an engineering team to consider people and organizational issues in the design of a system. As a solution, HSI is a set of properties, behaviors, structures, and functions that enable end users (i.e., people and organizations that are using the system).

HSI programs are based on the integration of human needs, skills, and capabilities from the beginning. Such programs contribute to reduce systems' life cycle costs and increase systems' efficiency, usability, and quality. They also contribute to reduce

1. Risk of error due to inadequate information (i.e., poorly designed user interfaces)
2. Risk associated with poor task design (i.e., insufficient or poorly conducted task analysis)
3. Risk of reduced safety and efficiency due to poor human-centered design (HCD) (i.e., poor integration of human factors in the design process) (Fitts et al., 2008; Sheridan, 1984).

---

[1] INCOSE HSI Working Group definition of Human–Systems Integration (HSI) is the following: "HSI is an interdisciplinary technical and management process for integrating human and organizational considerations with and across all system elements, an essential enabler to systems engineering practice. Human activity considered by HSI includes designing, modeling and simulation, engineering, operating, maintaining, supporting and dismantling the system. HSI also considers training and training devices, as well as the infrastructure used for operations and support (DAU, 2010). HSI incorporates the following domains as integration considerations: manpower, personnel, training, human-centered design, human factors engineering, occupational health, environment, safety, habitability, and human survivability."

Human-in-the-loop evaluation is always beneficial to the discovery of emerging behaviors and properties, because it enables evaluation of human's actions and organizational capacity and support. Evaluation metrics have to be defined, standardized, and further used within an organization. More specifically, human performance has to be measured using performance metrics, such as accuracy, workload, situation awareness, cognition, personality, emotional health, physiology, and biomechanics.

Systemic representation: structure and function | *human and machine* | *cognitive and physical*

HSI cannot be properly defined without defining what the concept of "system" really is about. The definition presented in this section breaks with traditional meaning of system conceived as a machine only but instead encapsulating humans, organizations, and machines.[2]

First of all, **a system is a representation** of

1. A human or more generally a natural entity (e.g., a bird, a vegetal)
2. An organization or a social group (e.g., a team, a community)
3. A machine or a technological entity (e.g., a car, a motorway).

A system is a system of systems, which means that a system includes people, machines, and systems. This recursive definition is crucial to be clearly understood. A system is multi-agent in the artificial intelligence (AI) sense; an agent being a society of agents.

A system has at least a structure and a function. In practice, it has several structures and several functions articulated within structures of structures and functions of functions. It is interesting to remind the analog definition of an agent in AI provided by Russell and Norvig (2010), which is defined as an architecture (i.e., structure) and a program (i.e., function).

A system has structures and functions that can be **physical and/or cognitive**. Figure 4.1 presents a synthetic view of what a system is about.

---

[2] This systemic view takes Herbert Simon's view of the *Science of the Artificial* (Simon, 1996), in the sense that he rejected to treat human sciences using the exclusive model of natural sciences (i.e., submission to natural laws) and to break between science and humanities by looking for a common ground that links them. The *Science of the Artificial* is looking for new constructs that would explain things, which were not understood before. These artificial constructs could be a language, an ontology, a conceptual model, or any kind of representation that makes sense.

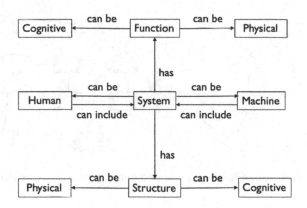

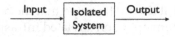

Figure 4.1 Synthetic view of the system representation.

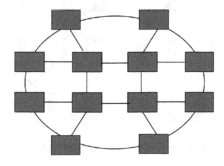

Figure 4.2 An isolated system.

For a long time, engineers used to think about a system as an isolated system, or a quasi-isolated system, which has an input and produces an output (Figure 4.2). As for an agent in AI, which has sensors and actuators, a system has sensors to acquire an input and actuators to produce an output.

However, as for an agent in AI, which is defined as an agency of agents (Minsky, 1986), a system is defined as a system of systems. Each system should be interconnected to other systems either statically (in terms of systems' structures) and/or dynamically (in terms of time, systems' functions, and more specifically, allocation of functions among a network of systems or system of systems). Summarizing, a system, as a system of systems, should be represented by a network of systems (Figure 4.3) where a network of functions could be dynamically allocated.

Figure 4.3 A system of systems represented as a structure of structures.

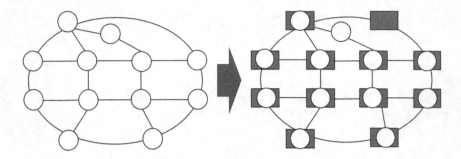

Figure 4.4 A function of functions mapped onto a structure of structures.

A system of systems is projected onto a structure of structures, usually called an infrastructure. Typically, functions are allocated to this infrastructure (i.e., a function of functions mapped onto the structure of structures – Figure 4.4). It should be noted that the network of functions is not necessarily a direct mapping on a related infrastructure.

Emergent behaviors and properties

We need to make a distinction between deliberately established functions allocated onto an infrastructure and functions that necessarily emerge from system activity. Indeed, systems within a bigger system (i.e., a system of systems) interact among each other to generate an activity. Bertalanffy (1968)[3] said: "A System is a set of elements in interaction." Emerging functions are discovered from such an activity (Figure 4.5).

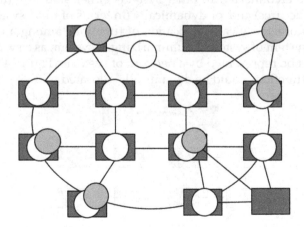

Figure 4.5 Emerging functions and structures within an active system of systems.

[3] www.sebokwiki.org/wiki/What_is_a_System%3F.

The integration of such emergent functions into the system of systems may lead to the generation of additional structures, which we also call emerging structures.

Systems of systems properties

A system's **purpose** is mainly defined by its task space (i.e., all tasks the system can perform successfully). Each task is performed by the system using a specific function that produces an activity that can be fully or partially observable (Figure 4.6).

A system's **functions** include perception of environment, action on this environment, and inference of actions from percepts. These functions can be physical and/or cognitive. Since systems interact among each other, some of their functions are socio-cognitive.

System's functions are recursive. Indeed, they are functions of functions as already defined. A system's function is teleologically defined by three entities (Figure 4.7):

1. Its role within the system of systems
2. Its context of validity that frames the boundaries of system's performance

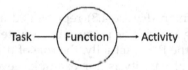

Task ——(Function)——→ Activity

Figure 4.6 A function logically transforms a task into an activity.

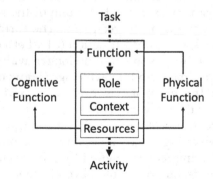

Figure 4.7 A function teleologically defined by its role (related to a task to be performed), context of validity, and resources, which can be physical and/or cognitive, producing an activity.

Figure 4.8 Postman.

3. Its set of resources required to perform its role with its context of validity. Resources are systems themselves that have their own cognitive and/or physical functions.

Let's consider a postman (Figure 4.8) represented as a system with the function of delivering letters. The postman as a system is part of a system of systems, which is the Post. Actually, the role of this system is "delivering letters." Context of validity is, for example, seven hours a day, five days a week (i.e., time-wise context in France, for example), and a given neighborhood (i.e., space-wise context). Resources can be physical (e.g., a bicycle and a big bag) and cognitive (e.g., a pattern matching algorithm that enables the postman to match the name of the street, the number on the door, and the name of the recipient). The corresponding pattern matching algorithm is a cognitive function (CF). Let's consider now that there is a strike and most postmen are no longer available for delivering letters. Remaining postmen should have longer hours of work and bigger neighborhood, until this expansion is too extreme and postman's helpers are needed to achieve the delivery task.

In this case, a tenure postman should have cognitive resources such as "training," "supervising," and "assessing" temporary personnel. We see that a CF "delivering letters" owned by a postman (i.e., an agent or a system) has to be decomposed into several other functions allocated to temporary postmen. We start to see an organization developed as an answer to a strike. More generally, **a function of functions can be distributed among a structure of structures**.

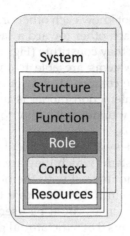

Figure 4.9 System's functional recursivity.

More specifically, function's **resources are systems** that have at least a structure and a function. This recursive property can be represented as in Figure 4.9.

System's **performance** is measured with respect to metrics, measurement models, and observable variables, which constitute data that should be analyzed and transformed into information for the construction of relevant knowledge on system's physical and cognitive attributes.

System's **structure** may vary a lot with respect to efficiency, compactness, and flexibility. It depends on the nature of the environment and the level of detail of its decomposition into sub-systems.

Systemic interaction models and system's capabilities

There are three main kinds of **systemic interaction models**:

1. Supervision
2. Mediation
3. Cooperation.

This threefold concept of interaction among agents was developed in the field of human–computer interaction (HCI) (Boy, 2011). In this book, it is extended to HSI.

Supervision is when a system (i.e., a supervisor) supervises interactions among systems. Supervision is about coordination. This interaction model is used when systems do not know each other or have enough resources to properly interact among each other toward a satisfactory performance of the system of systems that they constitute (Figure 4.10).

Mediation is when systems are able to interact among each other through a mediation space composed of a set of mediating systems (i.e., like ambassadors). This interaction model is used when systems barely know each other but easily understand how to use the mediation space (Figure 4.11).

Cooperation is when systems are able to have a socio-cognitive model of the system of systems which they are part of. Each system uses its socio-cognitive model to interact with the other systems to maximize some kinds of performance metrics. Note that this principle is collective and democratic. Other principles could be used such as dominance of a system over the other systems (i.e., a dictatorial principle). This interaction model is used when systems know each other through their own socio-cognitive model, which is able to adapt through learning from positive and negative interactions (Figure 4.12).

Knowing that systems are representations of humans and machines, it is clear that cooperation is mostly a capability of humans. However, AI

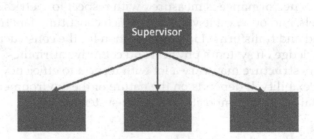

Figure 4.10 Supervision of systems by a system.

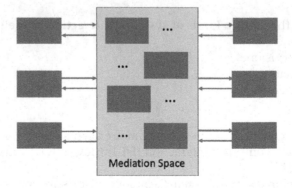

Figure 4.11 Mediation among systems through a mediation space.

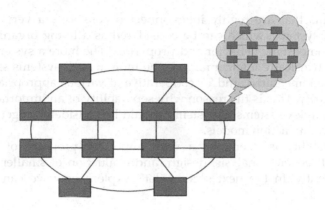

Figure 4.12 Systems cooperating among each other thanks to their knowledge of the others.

brings new ways of providing machines with such cooperation capabilities. First, AI provides situation awareness, decision-making, and planning capabilities in specific contexts. In addition, AI can also provide machines with learning capabilities and, more specifically, possibility of upgrading system's socio-cognitive model from experience, again in specific contexts.

The separability property

The term of separability has been used by physiologists to denote the possibility of separating an organ from the human body to work on it separately and putting it back. Some organs (i.e., systems) are separable, that is, the overall body (i.e., a system of systems) does not die from this momentary separation (Figure 4.13). Some other organs cannot be separated because the human being could die from this separation. Therefore, those organs have to be investigated and treated while connected to the rest of the body.

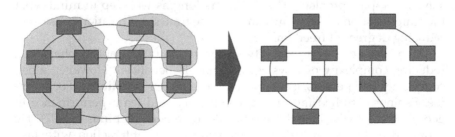

Figure 4.13 In the system of systems on the left, four sub-systems of systems are separable and are separated on the right for further processing.

Systems that are highly interconnected constitute a very complex system of systems, which can be considered as a living organism with its own homeostatic behavior and properties. The more a system of systems is intrinsically interconnected, the more its sub-systems should be coordinated, mediated, and/or cooperative at various appropriate levels of granularity. This is the reason why separability of an important property of complex systems of systems should be considered together with embedded interaction models.

Separability is, therefore, a very important property of complex systems. It enables analysis, design, and evaluation of smaller systems independently. In the next section, let's explore how we can manage complexity.

Complexity management

René Thom wrote for nonmathematicians an introduction to a very general categorization of form changes, which he called "catastrophes" (Thom, 1972). In a set of observables representing local states of a system, Thom defined a closed subset of catastrophes, in which the local phenomenological type of the system does not change as long as the representative point of the system does not meet the closed subset of catastrophes. When it meets it, there is a discontinuity in the appearance of the system, that is, interpreted as a morphogenesis.[4] Applying differential topology methods, Thom distinguished mathematical models adapted to regular processes from those that enable the description of the morphogenesis and analyzed the bifurcation phenomena that "generates the catastrophe." This essay is deeply original and marks the birth of catastrophe theory that experienced many developments since then.

More specifically, Thom developed catastrophe theory to explain that reducing degrees of freedom of a system may end up in nonlinear bifurcations that he called "catastrophes." For example, the projection of a 3D object on a 2D surface is likely to introduce bifurcations and then catastrophes. In engineering, it is common that we simplify in order to solve a complex problem. This is fine as long as we keep in mind what the simplification is about since most of the time simplification consists in reducing degrees of freedom.

Considering the separability property and the catastrophe issue, reducing complexity of a system is certainly beneficial. However, when complexity still persists, it is important to consider **familiarity**. For example, Espinosa studied familiarity, complexity, and team performance in geographically distributed software development (Espinosa, 2007). He provided a theory on task and team familiarity interaction with task

---

[4] Morphogenesis is the biological process that causes an organism to develop its shape.

and team coordination complexity to influence team performance. "Task familiarity is more beneficial with more complex tasks (i.e., tasks that are larger or with more complex structures) and that team familiarity is more beneficial when team coordination is more difficult (i.e., for larger or geographically dispersed teams)." More specifically, he found that task familiarity improves team performance when team familiarity is weak and vice versa.

Therefore, complexity can be managed more effectively when people become more familiar with systems being handled. More generally, familiarity should be distributed among systems of a system of systems. That obviously requires that we define in more detail what familiarity means. Familiarity is a matter of awareness and experience, which involves learning. Formalizing the incremental construction of the familiarity function between two systems, knowing that a system is a system of systems, familiarity supports activities of both Systems 1 and 2, which themselves provide inputs for learning processes feeding awareness and experience of both systems in terms of familiarity (Figure 4.14).

Familiarity is improved by standardization of tasks, harmonization of structures and functions, usability of technological tools, and improvement of organizations tools. Organizations tools, such as Six Sigma, tend to improve capability of business processes (Cano, Moguerza & Redchuk, 2012). This increase in performance and decrease in process variation helps lead to defect reduction and improvement in profits, employee morale, and quality of products or services. The main problem with **standardization** is possibility of blocking creativity, and therefore innovation, which could be very useful in cases of unexpected events. We can see that there is always an optimization between standardization and innovation. This is the reason why we need to develop standards that enable safer, more efficient, and comfortable operations, and potential flexibility in abnormal and emergency situations.

Engineering students learn to simplify a problem in order to solve it. In other words, they learn to reduce the problem space to a reasonable

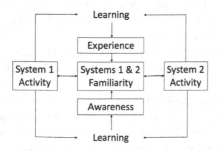

Figure 4.14 Construction and use of familiarity function.

solvable problem (usually a problem that they know solve). Most of the time, this simplification or reduction consists in eliminating nonlinearities. **Reductionism** can be represented as a projection of n-dimension space to a p-dimension space, where p < n. We already know that doing this leads to (mathematical) catastrophes, in Thom's sense (Thom, 1972). More generally, reductionism leads to unexpected situations. This is the reason why complexity of a system being designed is a crucial issue that deserves lots of attention. Consequently, in life-critical systems, it is much better to become familiar with a complex system instead of simplifying it.

Integrate as early as possible

Too many times, even if the various parts of a system are very well done, industry performs technological integration very late in the development process. This kind of practice results in maturity issues. In addition, HSI is even done later by developing user interfaces and operation procedures. Instead, HCD promotes integration, and more specifically HSI, at design time. We will see in Chapter 6 that creativity is integration. Indeed, integration is a matter of exploration of possible combinations of the various parts of a system. Such exploration involves tests. Integration requires insight, disciplined exploration, and test capabilities, which are not necessarily what engineering is about these days unfortunately. **Integration starts at design time, when high-level requirements are stated**.

Engineers are trained to be technicians and executants of procedures and methods that they apply to problems that are not necessarily well stated. Why? Primarily because our private and public technocratic structures dictate what should be done for financial profit purposes and not for human-centered purposes. But this is not the only cause, when high-level technocratic management is not enough knowledgeable in the sociotechnical aspects of the technology being developed, it may result in providing directives that are not tangible. This is the reason why start-up company's management is usually much more efficient than large companies, just because designers, engineers, business managers, and end users can be involved in a **participatory design** process from Day 1 to whatever it takes to deliver a good product. They are **goal-driven** and integration is usually addressed straightaway.

In addition, conventional engineering of the 20th century produced technicians who were and still are focused on a specific field or discipline and not on integration of complex systems. They are employed for executing very specific tasks and do not have an overall vision of the system of systems they are involved in (and is the ultimate goal!). Very few have the whole picture. This is what HSI is about: "**the whole picture!**" HSI has to consider interconnected systems of systems, where systems include people and machines. This kind of requirement does not fit the current urgency of fast money making... and fast everything! It takes time to get complex systems working fine and becoming mature technologically, organizationally, and humanly. This is the time it takes to imagine possible futures for such complex systems, exploring various kinds of usages and testing them. It is often much better to perform these **projection, exploration,** and **tests** during the design and development phases of a system than compensating design flaws by a lot of costly repair solutions once the system is delivered and used.

Integration, more specifically HSI, contributes the optimization of time to market by improving agility of the development process, alignment with job requirements, and operation's effectivity in terms of time and content satisfaction. Operation costs will be reduced by insuring system coherence, increasing tangibility, and automating the value chain. Productivity will be increased by rethinking services and organizing teams sourcing by increasing application productivity. However, integration is always a matter of context.

How can we define and construct context?

The concept of context could be defined and described in a lot of ways. Context is usually related to entities such as time, situation, behavior, point of view, relationships among agents, discourse, and dialogue. In practice, context is defined as patterns of less complex context patterns. This recursive definition will be detailed in the following sections. Context patterns can be seen as cognitive filters used to describe people's activity. They combine situational conditions that are purposefully matched to the current perceived situation to identify it and suggest appropriate actions. In many cases, it is practical to use the following distinction: verbal, social, or historical (Boy, 1998):

- **Verbal context**, that is, what is around the object and defines it in some situations, such as a word or phrase, a picture, a movie sequence, a physical object (such as a table), or a person.
- **Social context**, that is, what one can point out in the environment of the object to concretize its definition, such as time, the type of agent, or the type of task an agent is currently doing. Characterizing social context, Mantovani (1996) distinguished *physical context* from *conceptual context* by saying that concepts framing actions are grounded in the individual's sense of place, role, and value; I do not make this distinction in this book, and I consider that social context includes both physical and conceptual contexts.

- **Historical or temporal context** is essential in dynamic systems. It can be a dynamic window which shows the state of the environment including the various agents involved (e.g., their intentions, focus of attention, perceived state of their environment). The use of dynamic systems provides a very large number of situation categories. Expert agents index these situation categories with respect to context. This indexing enables agents to keep an anticipated awareness of the temporal progression. Historical context can also be described as what happened before and is causally related to the object being described.

As Hayakawa (1990) pointed out, words have *extensional* and *intensional* meaning. If a word can be described by a real object that you can see in the environment, its meaning is extensional. For instance, if you point at a cow in a field, you denote the cow. If a word is described by other words to suggest its presence, then its meaning is intensional. For example, you may describe a cow by saying that it is an animal, it has horns, it lives on a farm, etc. Each object can be associated to a particular intensional description according to who is giving the description, when the description is given, etc. Again, context is a key issue when objects have to be described. It is a fallacy to claim that each object could be associated with a single word. Anyone who has tried to retrieve documents in a library using keywords equations knows this. It is almost impossible to retrieve the desired information because the description one given for the query seldom matches the descriptions that librarians have developed (verbal context). However, if you ask the librarian, he or she may help you better by acquiring more (social) context from you. In the best case, if the librarian is a good friend and knows your work and needs, then he or she will be very helpful. If the librarian does not know you, he or she can capture social context by simply looking at you. He or she may consider facts such as "you are young" and "you wear a lab coat." He or she will also capture context from what you say (verbal context). Hayakawa says that "an examination of the verbal context of an utterance, as well as an examination of the utterance itself, directs us to its intensional meanings; an examination of the physical [social] context directs us to the extensional meanings."

Context includes a temporal aspect, that is, context summarizes a time period. Furthermore, persistence of some events reinforces context. For instance, a young scientist told the librarian that she is looking for some information on geometry, she is currently involved in a computer class, she has a problem of drawing curves using a set of points, she needs to obtain a continuous drawing on the screen, and later, she finally requested a reference on splines. The librarian usually integrates *historical context* and, in this specific example, will not be confused between mathematical spline functions (what the requester needs) and the physical splines that draughtsmen used in "ancient" times. In this example, the words "computer" and "screen" help to decide that the requested "splines" will be used in the last quarter of the 20th century.

The concept of context is multidimensional

The development of a function analysis is based on the teleological definition of functions (see Figure 4.9). One of the major factors is "context of validity" of physical and/or CFs being developed and analyzed. This section is devoted to a deeper analysis of what context is about in HCD.

Context is a very difficult concept to grasp. It is related to several concepts, such as situation, organization, culture, location, time, behavior, point of view, relationships among agents, discourse, and dialogue. The difficulty that we usually have to define "context" mainly depends on the viewpoint that we have (e.g., whether being an engineer making a system or a human operator using the system). The following distinctions provide a framework that supports context definition (Figure 4.15):

1. Cognition/physical (i.e., what is generated by people and what is not)
2. Design/operations (i.e., functioning logic versus use logic)
3. Normal/abnormal (i.e., normal versus abnormal situations).

Let's start with the **cognition/physical distinction**. We already grasped what the AUTOS pyramid is about and its usefulness in HCD (see Figures 3.2–3.4). It took me a while before I understood that "organization" should be considered independently from the "situation." Originally, I was talking about "organizational environment" by mixing "socio-cognitive organization" and "physical situation" (Boy, 1998).

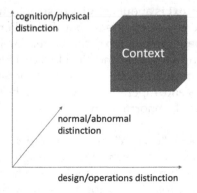

Figure 4.15 A three-dimensional framework for context definition.

Using CF analysis in various kinds of HCD projects, it became clear that designing an artifact that supports the execution of a task by a specific user requires consideration of two types of interaction: interaction with other agents in the socio-cognitive organization where the artifact will be used and interaction within various kinds of physical situations.

Organizational context can be investigated as a socio-cognitive topology (i.e., cognitive structures of structures). We will refer to the synthetic view of the system representation presented in Figure 4.1. An organization is an agency of agents. An organization is concerned with the properties of a structure that are preserved under various kinds of stresses but not in others where the organization should be changed. For example, an orchestra is an organization that includes musicians and various kinds of instruments (technology), which are topologically inter-related to produce a coherent symphony (i.e., the product). We can say that a musician plays in the context of a specific orchestra (i.e., he or she would not play the same in a different orchestra). Situational context can be investigated as a physical topology (i.e., physical structures of structures of the location where the symphony is being performed).

The **engineering design/operations distinction** is about designers and engineers' viewpoint of an artifact they are designing versus end users' and human operators' viewpoint of the same artifact they have to use or operate. This distinction is extremely useful in HCD since engineering design context and operations context are usually very different, and both designers and operators should have common understanding on ways artifacts being designed works and can be used.

- **Engineering design context** is about creativity and rationalization of sociotechnical systems leading to acceptable solutions. It is clear that the more creativity and rationalization are well supported at design time, the better the design team will make a good job.
- **Operations context** is about activity of sociotechnical systems when they are put at work. It is clear that the earlier a design team can understand operations context, the better for the requirements of the system being designed and developed. This is the reason why human-in-the-loop modeling and simulation are very useful support toward this end. Operations context can be split into three main categories: normal, abnormal, and emergency.

We typically talk about "usages" in the operations context. One of the main goals of HCD is to consider usages at design time. Conversely, it is also as important that human operators know why an artifact has been designed the way it is. Cross-fertilization of these two types of viewpoints always improves artifact design in the end. This is what participatory design is about.

We already presented the design card mechanism (Figures 1.9–1.11), which associates four processes: rationalization, activity, structure, and function. Rationalization describes the reasons why a system has been designed the way it is. Activity describes effective usages of the system (i.e., effective operations). Structure and function describe the system itself. Design cards not only capture design rationale and structural/functional description of the system but also the way it is used. Design card cannot be fully informed without Human-In-The-Loop Simulation (HITLS) that provides means for activity analysis. Unlike task analysis that is typically based on a priori knowledge, activity analysis is based on observation of the activity of the system being documented. Never forget that HITLS requires tangibility assessment (see Chapter 2).

Ultimate goal is integration of people and organizational setups at the system level. HSI provides a philosophy and methods that can be used during the whole life cycle of a system. HSI should be started from the beginning of the design phase in order to improve joint human–systems operations and anticipate downstream risks in cases of human errors and system failures. However, we still see in industry systems engineering considering machine systems first and people second. Even if cognitive engineering and HCD have been developed and used for more than three decades, HSI is still a challenge. This challenge will be discussed in the conclusion of this book, going beyond the design/operations distinction.

The **normal/abnormal distinction** is crucial. During the last three decades of the 20th century, we automated machines like never before. We also observed that automation rigidifies practices, standardizing them over procedures handled either by people (i.e., automation of people) or machines (i.e., automation of machines). However, automation is always designed within very normalized contexts of use. We talk about normal or nominal conditions or situations. Outside of these normal contexts, automation can be counterproductive and even catastrophic. Today, when we say "automation," we can also say "machine systems" or "systems" where people and organizations are included. In abnormal situations, automation involves operational rigidity – at the time we actually need flexibility the most. Why? When everything falls apart, we need to state and solve problems correctly. In some well-known abnormal and even emergency situations, operations procedures are available – they should be used. In aeronautics, for example, we have abnormal and emergency procedures for a large number of situations. They are very effective. However, there are other situations that cannot be anticipated well enough in advance, and therefore, operations procedures cannot be produced for those situations. We talk about unexpected events or situations.

Persistence and mutual inclusion of contexts

Taking an HCD perspective, context should be defined using three scientific perspectives originating from social sciences (e.g., sociology, anthropology, ethnology), psychology, and life science. I claim that these perspectives lead to constraints that are mutually inclusive (i.e., social constraints include psychological constraints that include biological constraints).

Mantovani proposed three mutually inclusive[5] levels of social context definition and construction (Mantovani, 1996):

- Level 1: **context construction**, that is, the social context is determined by *values* that are very general goals determined by the *culture* (Thomas & Alaphilippe, 1993). This level associates culture, context, and history.
- Level 2: **(everyday) situation interpretation**, that is, the psychological level, motivations are more precise but less persistent than attitudes and are influenced by values and needs. This level associates opportunities, interests, and goals.
- Level 3: **local interaction with the environment**, that is, for human agents, this is the biological level and the needs are biological strengths such as eating, drinking, or sleeping. This level corresponds to the AUT triangle (artifact, user, and task) described in Figure 3.2.

Mantovani's three-level model can be interpreted as follows (Figure 10.2):

- A user may have situated interests for action in a given context
- The task that a user performs is based on situated goals coming from a social history
- The tool is built from opportunities that arise from a cultural structure.

Each task context is included in another task context that is more *persistent*. The town is more persistent than the building, the building is more persistent than the room. The phase is more persistent than the sub-phase, and so on. Each context attribute is included in a more persistent context attribute.

Context can be organized into *context islands* with respect to its attributes and the persistence of these attributes. Context islands may be

- *Hierarchically dependent*, they are then defined as mutually inclusive context patterns with respect to their relevant attributes and the degree of persistence of these attributes

[5] That is Level 1 is more general than Level 2 which is more general than Level 3.

- *Independent*, they are then defined as mutually exclusive context patterns with respect to their relevant attributes and the degree of persistence of these attributes
- *Interdependent*, they then share some context patterns.

Dealing with system's dynamics

The third distinction is related to dynamic nature of the system being studied. Three categories of dynamic systems can be distinguished and instantiated by the coffee maker, the car, and the airplane (Boy, 1998). You can stop your coffee maker if there is a problem. Nothing dramatic will result. If you stop your car for any reason, let's say to avoid a pedestrian crossing the road, you might cause another accident by the fact that the car following you did not anticipate such an unexpected event. In this case, you can stop but… If you are a pilot and you are facing a severe problem on board, you just cannot stop the airplane, otherwise you fall! In this case, you cannot stop at all.

- **Unconstrained dynamic systems**, such as the coffee maker, can be stopped safely at any time independent of the current evolution of the environment. They are evolving with time in an open-loop fashion. In other words, you can anticipate the final conditions before stopping.
- **Loosely constrained dynamic systems**, such as the car, can be stopped according to conditions set by the environment. They are evolving with time in a closed loop with the evolution of their environment. In this case, human operators have to physically be in the loop. In other words, you cannot fully anticipate the final conditions before stopping.
- **Strongly constrained dynamic systems**, such as the airplane, cannot be stopped at any time when in operation. They are evolving in a closed-loop fashion. Beside this closed-loop evolution, the physical machine or even the physical world has become too distant from human operators.

Knowledge involved in failure recovery depends on the type of dynamic system. The first type is deterministic. Thus, the dynamic system can be operated without requiring attention. Even if the operator makes an error, the result in the manipulation of the corresponding tool will not cause any dramatic problems. In the second type, human activity requires attention during operation. Human operators have to compromise between choices. Knowledge about such compromises is difficult to elicit. Activity variation around the task requirements is difficult to predict or explain offline. Thus, observation methods must be used, and reporting systems are frequently implemented. In the third type, human activity demands continuous attention. The machine is a reactive agent. Cooperation between the human and the machine governs the entire stability of the overall system. Elicitation of this cooperation knowledge involved in the control task is extremely difficult for knowledge engineers that are novices in the expertise domain. Both self-training and frequent observation of real experts are always necessary.

Dealing with the unknown

Unexpected situations can come from intrinsic situations, themselves coming from system design (technology), organizational setups or people's activities (the technology, organizations, and people (TOP) model again!), and from extrinsic situations (e.g., environmental events difficult to anticipate). We can develop a body of knowledge on unexpected intrinsic situations by using HITLS and experience feedback mechanisms and, therefore, use it to incrementally fine-tune the overall system.

Normal, abnormal, and emergency contexts can be known or unknown (Figure 4.16). In the known space, domain knowledge can be specified and documented for further use. In the unknown space, problem solving is at stake based on related knowledge and experience of people involved at operations time. We see here the importance of experience feedback for skills and knowledge reuse.

Combining physics, cognition, philosophy, and design

Let me tell you an interesting story. For the last few months, I realized that my computer and its charger were heating up disproportionately. I first thought that this was a hardware issue... the battery perhaps!

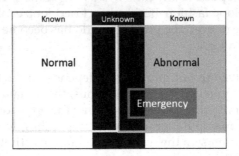

Figure 4.16 Context distinctions: normal, abnormal, and emergency versus known/unknown.

My computer was getting horribly warm erratically (at least this is what I thought). I went to the store, and someone very competent immediately found the cause of my troubles. He verified the energy consumption for all applications on my computer and found that an application was far ahead of the others in terms of energy consumption, and he turned the hardware accelerator switch off. As a matter of fact, this application is known as the fastest software of its category. What was not said is that this software speed advantage had a cost in terms of energy consumption. In order to be that fast, it has to use more central processing unit (CPU) resource than other applications of that type are willing to use; and more CPU usage means more battery drainage. This is a good example of a strong relationship between software and hardware and between information and energy.

This information–energy association does not go without appropriate philosophical constructs and practical implementation means. Designing systems for people requires explaining correctly what they are about. We need models that provide interoperable concepts and the relationships among them. This is the reason why I developed two articulated models, NAIR and SFAC (Boy, 2017), which are restated below in order to better understand HCD foundations of tangible interactive systems.

The NAIR model: natural/artificial versus cognitive/physical

The NAIR model (natural/artificial versus cognitive/physical) was developed to rationalize the cognitive-physical functional distinction from both a philosophical point of view (i.e., the natural side) and an engineering design point of view (i.e., the artificial side). Cognition has to do with intentional behavior supported by rationalist[6] philosophy on the natural side and AI on the artificial side and physics with reactive behavior supported by vitalist[7] philosophy on the natural side and automatic control/HCI on the artificial side (Figure 4.17).

Artificial (hardware- and software-based) or natural systems have functions that are either physical or cognitive. Natural systems include, for example, people, animals, plants, and physical systems (e.g., geologic or atmospheric phenomena). Artificial systems include, for example,

---

[6] Rationalist philosophy is mainly related to the cortex, including reasoning, understanding, and learning (Markie, 2013).

[7] Vitalist philosophy was developed by Henri Bergson, mainly related to the reptilian brain, including emotions, experience, and skills (Bergson, 1907). It is also related to Nietzsche's "will to power" concept that was close to Schopenhauer's "will to live," a psychological force consciously and unconsciously used to survive (Wicks, 2011).

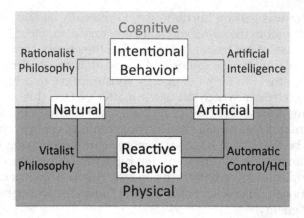

Figure 4.17 The NAIR model: cognitive (intentional) and physical (reactive) functions (Boy, 2017).

information technology (e.g., aircraft flight management systems, Internet) and mechanical systems (e.g., old mechanical watches[8]).

Let me take an example. "Planning a trip" is a complex CF, which involves other CFs such as "defining a budget," "choosing transportation means," and "defining time constraints." Each of these CFs involves smaller grain CFs (i.e., function of functions). The concept of CF is, therefore, recursive (i.e., a CF is an organized set of CFs). However, it may happen that a CF requires the use of a physical tool involving physical functions, such as "using paper and pencil to write down an itinerary." It is practical to extend both concepts of CF and physical function to the generic concept of **resource**. A resource is a system. We will then say that a resource can be physical or cognitive.

The SFAC model

Designing a system is defining its structure and function. Each structure and function can be described in an abstract way and a concrete way. The SFAC model (structure/function versus abstract/concrete) provides double articulation (i.e., abstract and concrete) between system structure and function (Figure 4.18) as follows:

- Declarative knowledge (DK) (i.e., abstract structures)
- Procedural knowledge (PK) (i.e., abstract functions)
- Static objects (i.e., concrete structures)
- Dynamic processes (i.e., concrete functions).

---

[8] Note that current watches include both software and hardware.

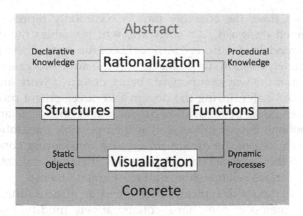

Figure 4.18 The SFAC model (Boy, 2017).

The abstract part is a rationalization of the system being designed (i.e., knowledge representation). This rationalization can be represented by a set of concepts related among each other by typed relationships. This kind of representation can be called ontology,[9] semantic network, or concept map. It can take the form of a tree hierarchy in the simplest case or a complex concept graph in most cases. Ontology use in HCD will developed in Chapter 2.

The "declarative" and "procedural" terms,[10] respectively, refer to knowing what and knowing how. They are used to describe human memory. Declarative memory includes facts and defines our own semantics of things. Procedural memory includes skills and procedures (i.e., how to do things). We can think declarative memory as an explicit network of concepts. Procedural memory could be thought as an implicit set of know-hows. Declarative memory and procedural memory are both in the cortex and involve learning. The former is typically stored in the temporal cortex of the brain. The latter is stored in the motor cortex.

The distinction between DK and PK was used in AI for a long time, especially in the expert system domain. DK includes factual and fundamentally static information that describes how things are and can be related among each other. PK is about how to perform and operate. It can be called know-how or skills.

---

[9] In philosophy, ontology is the study of what there is, what exists. It is "what the most general features and relations of these things are" (Hofweber, 2018).

[10] These terms are used in computer programming that began by being thought and implemented in the procedural way. For example, languages such as Fortran were developed on the basis of subroutines, and Pascal used procedures. Subroutines and procedures enable programmers to develop procedural knowledge. Then AI came and proposed declarative programming, such as Lisp and Prolog (i.e., defining objects, functions, predicates, and methods).

At design time, the concrete part is commonly represented using computer-aided design (CAD) software, which enables the designer to generate 3D models of various components of the system being designed. These 3D models include static objects and dynamic processes that allow visualization of the way components being designed work and are integrated together. Later during the design and development process, these 3D models can be 3D printed, allowing for a more graspable appreciation of the components being built as well as their possible integration. Testing occurs at each step of the design process by considering concrete parts together with their abstract counterparts (i.e., their rationalization, justifications, and various relationships that exist among them).

The SFAC model is typically developed as a mediating space that design team members can share, collaboratively modify, and validate. SFAC also enables the design team to support **documentation of the design process and its solutions** (see Chapter 1; and Boy, 1997). The concept of active design document (ADD), initially developed for traceability purposes, is useful for rationalization of innovative concepts and incremental formative evaluations (Boy, 2005). The SFAC model was the basis of the SCORE system used to support a team, designing a light water nuclear reactor (see Figure 1.10 and explanations around it), in their collaborative work and project management (Boy et al., 2016).

Visualization is crucial in HCD because it is important that all stakeholders of a design process from its inception to development be aware of what the system being designed is about. Visualization is about tangibility (i.e., physical tangibility coming realistic visual representation of the system being designed at all stages, and figurative tangibility figuring out meaning of the system as a whole and its attributes).

Summary

This chapter is the central part of the book (Figure 4.19). First, formalizing HSI is about considering systems as multi-agent (i.e., systems of systems). Therefore, a system is interconnected both externally and internally. It is a representation of a human, a machine, or an organization. It can be cognitive and/or physical. It can be deliberately constructed or emergent from the activity of the system of systems it is part of or its own internal activity. It has a structure of structures and a function of functions. It has an input, called a task, and an output, called an activity. Formalizing HSI requires support of systemic interaction models (i.e., supervision, mediation, and cooperation models). Formalizing HSI requires understanding separability, complexity management, and context. Context deserves a lot of attention in HCD. It is multidimensional leading to several distinctions, such as cognition/physical that requires models such as NAIR and SFAC, design/operations, and normal/abnormal.

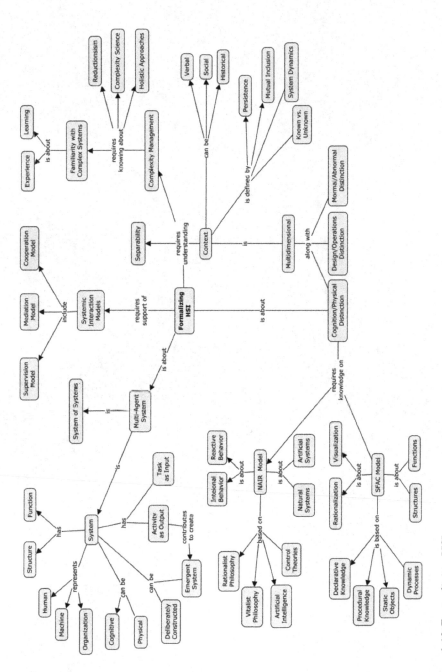

Figure 4.19 Formalizing HSI.

chapter five

From rigid automation to flexible autonomy

> While reductionists want to reduce mind to the brain, system scientists, such as Noam Chomsky, Humberto Maturana, Francisco Varela, John Holland and Gregory Bateson treat mind as a level of analysis unto itself: while the mind is, in part, a form of brain-based cognition, it is a complex system unto itself and therefore more than brain.
>
> **Brian Castellani and Frederic William Hafferty (2009)**[1]

For a long time, humans incrementally designed and developed tools and prostheses that expand their capabilities. Automation is one of these tools and prostheses. Reductionism brought automation where it is today. Even if it often helps tremendously, automation sometimes introduces **rigidity** in our lives. In addition, such rigidity is often experienced when things go wrong and sometimes tend to amplify to the point of contributing to real disasters.

In this book, automation is seen as cognitive function transfer from a human to a machine. For example, an aircraft autopilot controls speed and altitude and has cognitive functions dedicated to these kinds of tasks, in the same way a pilot would humanly control the aircraft. Of course, even if corresponding cognitive functions are similar and have the same goals, they are exactly run the same in a "human system" and in a "machine system."

Evolution of automation (Rasmussen's model)

Jens Rasmussen proposed and developed a model of human behavior in a process control situation (Rasmussen, 1986). This model is presented on the left side of Figure 5.1. It includes three behavioral levels, starting from the bottom:

1. **Skill**-based behavior that consists in a stimulus-response process or subconscious function (i.e., from information captured from system's environment through receptors, using operations rules, the human

---

[1] Sociologists and writers in medicine sociology and complexity science.

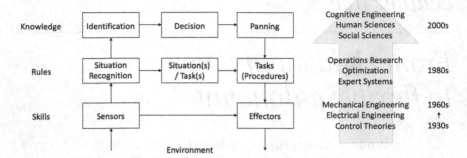

Figure 5.1 Automation evolution and emergence of contributing disciplines based on Rasmussen's model.

operator acts through his/her effectors on the environment, which may include other agents);

2. **Rule**-based behavior that consists in a conscious process or function (i.e., when the skill level does not work, the rule level is solicited to think about a procedure to apply as a function inferred from a situation recognized through a rule);

3. **Knowledge**-based behavior that consists in a conscious "highly cognitive" process or function, which includes three main functions: identification of the situation (i.e., provision of meaning to a situation not recognized at the rule level), decision-making (i.e., foresee possible futures and choose one of them with respect to a purpose or goal), and planning (i.e., establish a plan of actions that leads to an operations procedure).

Automation evolution and emergent contributing disciplines with respect to behavioral levels of Rasmussen's model are presented on the right side of Figure 5.1. At the skill-based level, automation involves disciplines such as electrical engineering, mechanical engineering, and control theories. Autopilots have been developed since the 1930s. Since the sixties, we really master this kind of automation both theoretically and practically. Skill-based automation feedback loop has an average time constant of 500 milliseconds.

At the rule-based level, automation involves disciplines such as operational research, optimization, and expert systems (i.e., AI rule-based systems). Flight management systems (FMSs) were developed in the mid-eighties on top of autopilots. The FMS enables pilots to plan for a flight and program the route using a database of flight routes. It is also capable of managing the flight of an aircraft from an airport to another, via a series of waypoints (i.e., after takeoff, the aircraft is capable of flying by itself if conditions remain the same as when the flight plan has been established). Rule-based automation feedback loop has an average time constant of 15 seconds to a minute.

At the knowledge-based level, automation is still not done. We believed that we could have AI algorithms capable of making sense of a situation especially in close, very specific, and limited contexts (i.e., where we practically know everything and there is no risk for people). Today, most contexts are open and legal issues are still fuzzy, and therefore, only people can handle the knowledge-based level at the moment. This is the reason why cognitive engineering[2] was developed based on cognitive science results, and human and social sciences. Knowledge-based level automation has an average time constant of several seconds to several minutes depending on the problem to be solved.

Designing for flexibility using the TOP model

Whether at skill or rule level, automation is implemented in the form of rigid procedures. Inevitably, when an unexpected situation occurs (i.e., a situation outside of the context of validity of the automation cognitive function), automation may act inappropriately (i.e., automation does not have any adaptation mechanism outside of its domain of validity). Skill- and rule-based automation works perfectly in normal situations. It is, therefore, interesting to better understand what kinds of tools would be suitable for expanding the knowledge-based level outside human operators' heads. In other words, what kind of performance support would be suitable in abnormal and emergency situations. **These tools should enable flexibility.**

Control and management of life-critical systems are typically supported by operations procedures and automation. Automation is usually thought as automation of machine functions. Analogously, operation procedures can be thought as automation of people (Boy & Schmitt, 2012). Problems come when unexpected situations occur, and rigid assistance (i.e., procedures and automation) does not work any longer, because we are out of its validity context. **Problem solving** is at stake and, more specifically, human problem solving. Instead of automation, people need autonomy, first for themselves (i.e., human autonomy) and from machines (i.e., machine autonomy). We then face a multi-agent system, where agents incrementally become more autonomous through learning and, therefore, should be coordinated. Figure 5.2 presents these three options, which lead to the difficult problem of **function allocation.**

---

[2] Cognitive engineering was born in the eighties and took two decades to develop. Today, cognitive engineering is part of human-centered design.

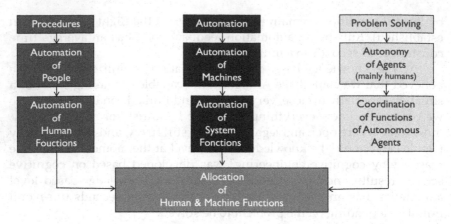

Figure 5.2 Procedures, automation, and problem solving leading to allocation of human and machine functions.

Solving a problem requires enough technological, organizational, and/or human flexibility (i.e., the technology, organizations, and people (TOP) model is back again!).

Technology-driven flexibility (i.e., the "T" of the TOP model)

Technology-driven flexibility is directly related to inventing the future and testing solutions. This kind of flexibility can be very well illustrated by the story of Engelbart's innovative constructs and novel information technology that he contributed to generate. Indeed, in the late 1950s–early 1960s, Douglas Engelbart was trying to "augment the human intellect" and developed a model that he called AUGMENT, and a demonstrator called NLS (oNLine System) (Engelbart, 1986). Engelbart's augmentation framework supported integration of computing technology, organization (development), and people (psychology)... the TOP model, one more time! Engelbart understood that we should look for symbiosis between people and computing machines through exploration of various kinds of computer applications. Engelbart's inventions[3] greatly contributed to the creation of the field of HCI.

---

[3] During his very "nonlinear" career, Douglas Engelbart was recognized as the inventor of the computer mouse and one of the most influential contributors to the concept of groupware and computer-supported collaboration, more specifically ARPANET (i.e., the precursor of the Internet), through the development of hypertext, networked computers, and precursors to graphical user interfaces (GUI). He was one of the first to demonstrate computerized teleconferencing. He developed a community of information architects working on ARPANET in the mid-seventies.

Engelbart's work was comparable to the work of Clément Ader, who invented, developed, and "flew" on the first flying machine, that he called Éole, in 1890,[4] and Orville and Wilbur Wright who invented, built, and made the first controlled and sustained flight of a powered airplane in 1903. Both types of machines (i.e., computers[5] and aircraft) are augmenting human capabilities, cognitively and physically. It is interesting to split such machines into two high categories: **tools** and **prostheses**. The former includes systems that extend human capabilities (e.g., a knife that enables people to cut something, knowing that human beings can cut these things in a suboptimal way without knives). The latter includes systems that enable people to do things that would be impossible to do without using them (e.g., aircraft that enables people to fly, knowing that human beings do not fly). Contemporary computers and, more importantly, the Internet are cognitive prostheses providing people with external information processing and memory capabilities.

Subsequently, Engelbart's interactive computing technology improved flexibility in text processing, mobility in easily using smartphones, and computer graphics handling, for example. Almost at the same time, artificial intelligence (AI) was born. John McCarthy and Marvin Minsky presented the first AI program at the Dartmouth Summer Research Project on Artificial Intelligence in 1956.[6] AI became big during the 1980s but did not go through until recently, leaving the field to human–computer interaction (HCI) during the 1990s and 2000s. We can say that HCI highly contributed to human-centered automation. It is hoped that AI of the 2020s will contribute to the development of **human-centered autonomy**, which should greatly provide people and machines with appropriate technological flexibility.

The area of augmented cognition developed in the beginning of the 2000s, associating AI, HCI, cognitive science, and human factors and ergonomics (HFE). The main idea is making the human–computer association more performant than the sum of each of them. Augmented cognition tries to put together augmented reality (AR), virtual reality (VR), and AI integrated within human-centered design (HCD). This includes systems

---

[4] It is recognized that Éole, equipped with a lightweight steam engine, took off in the vicinity of Paris in 1890, France, for an uncontrolled flight of approximately fifty meters at a height of approximately twenty centimeters.

[5] Of course, Douglas Engelbart did not invent the computer per se, but he gave birth to the kind of interaction we have with computers today. The British mathematician Charles Babbage designed between 1833 and 1871 the precursor of current computing machines. As a comparison, even if we had to wait until the 1940s to see a helicopter flying for real, Leonardo da Vinci sketched the first helicopter in the 15th century and can be credited for being a precursor of aviation also.

[6] We should also give credit to Allen Newell, Cliff Shaw, and Herbert Simon for anticipating AI with the Logic Theorist, a program designed to mimic the problem-solving skills of a human and was funded by Research and Development (RAND) Corporation.

equipped with tools such as computer vision, pattern recognition, head-mounted computing, and natural language understanding.

Organization-driven flexibility (i.e., the "O" of the TOP model)

Organization-driven flexibility is directly related to participatory design and collaborative operations. This kind of flexibility can be very well illustrated by the story of the Apollo 13's successful failure, as already described in the introduction of this book. Let's use corresponding experience feedback.

> The major strengths of NASA Apollo flight crews and mission control (i.e., ground personnel) were mutual trust and respect, and interdependence. Ground was considered an extension of crew and spacecraft. The team trained like they flew, and flew like they trained.
>
> *(Griffin, 2010)*

Space mission crew and ground personnel have to focus on situational awareness, teamwork, trust, disciplined execution, and, most importantly, do what they are trained for ("You've gotta fly like you train"). Post-flight operations were open and honest, conducting thorough debriefing. Emphasis was on problem areas. During the Apollo program, feedback was quick and corrective actions were implemented for the next flight.

Apollo 13's successful failure shows how a cohesive, well-trained, competent, and disciplined organization can overcome extremely difficult and dangerous situations. In other words, even if we have the best technology to support complex activities, operations will not succeed without a good organization well articulated with people and technology. Organizational flexibility is more specifically needed when things go wrong (i.e., in abnormal and emergency situations).

Figure 5.2 illustrates three kinds of operations process support: (1) procedural support, (2) automation support, and (3) problem-solving support. Supports 1 and 3 involve human functions: procedure following with the former (i.e., human operators need to follow procedures) and problem solving with the latter (i.e., they need to use their knowledge and human problem-solving capabilities). Supports 2 and 3 involve machine functions: automation (e.g., autopilots, FMSs, and advanced driver-assistance systems) and problem-solving specific tools (e.g., visualization, reporting coordination tools, collaborative systems). Supports 1 and 2 have been dominant up to now. They have been extremely useful in a large variety of situations. However, they are not very much useful in unexpected or unanticipated situations, where flexibility is needed to solve problems, sometimes urgently. Collaborative problem solving is, therefore, at stake and requires appropriate organizational culture, tools, and techniques, as what was already described for space programs such as the Apollo program.

Human-driven flexibility (i.e., the "P" of the TOP model)

Human-driven flexibility is directly related to competence. This kind of flexibility can be very well illustrated by the story of the heroic handling of DHL crew of the Airbus A300 B4 hit by a ground-air missile shot by Iraqi resistance in 2003 close to Baghdad. They successfully landed by adjusting engine thrust, thanks to their airmanship competence. Before we tell you a digest of the story and if you are not acquainted with aviation, it is necessary to provide you a few things on aeronautics knowledge.

Stability of an aircraft is one of the main issues that aerospace engineers have been working on since the beginning of aeronautics. Three modes are main ingredients of flying assets:

- Pitch (i.e., movement around aircraft transverse axis – axis parallel to the wings)
- Yaw (i.e., movement around aircraft normal axis – axis parallel to the fuselage)
- Roll (i.e., movement around aircraft longitudinal axis – the direction the pilot faces).

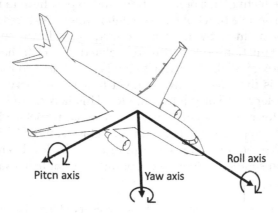

Figure 5.3 Aircraft pitch, yaw, and roll modes.

Pilots have to take care of these three modes. Figure 5.3 presents the three axes corresponding to these three modes.

Flying an aircraft is basically summarized in handling thrust and lift. When the pilot carefully trims pitch (i.e., adjusts downward lift of the horizontal tail) and flies hands-off, the aircraft flies straight for a while but then starts to cycle up and down in a slow mode: this is the phugoid phenomenon. Phugoid is a sinusoidal cycle that consists in putting nose down, gaining speed that causes increasing lift, aircraft starts climbing and loses speed, and so on.[7] This is precisely what DHL crew of the Airbus A300 B4 did prior to land in Baghdad in 2003. In this example, flexibility is coming from collaborative problem solving combined with individual skills and knowledge of human operators involved.

The need for function analysis

What should we do when we design a new system? We should determine what are the functions that need to be allocated to the various structures, whether machines or humans, whether physical or cognitive (refer to Figure 4.1).

From a human-centered point of view, function allocation should be done with respect to what Figure 5.2 presents (i.e., everything that should be automated should be allocated to machines, everything that can be formalized in the form of operations procedures should be allocated to people, and what remains should support problem solving). Typically, procedures and automation include what is really well known, mastered, and, if possible, contextualized (i.e., typically normal situations, as well as some abnormal and emergency situations). However, in some abnormal and emergency situations, operational support should be provided combining flexible technology, organization, and expert human resources.

How appropriate functions can be determined? As already described, a function is defined by its role, context of validity, and resources (Figure 4.12). Function's role is directly related to a task that needs to be executed. More generally, function's role is associated to a task category or type. This is the reason why we need to start by carrying out a task analysis in order to develop a task model. This task model can be represented in the form of a task ontology (i.e., a semantic network of task concepts related by typed links). For example, a task can be decomposed into sub-tasks and so on until a meaningful granularity level is reached. Such decomposition should be done deep enough to establish a successful task

---

[7] The fact the nose goes down when the aircraft loses speed is an important aspect of a stable aircraft. It means most aircraft will get themselves out of trouble (trouble here is losing so much speed the air stream breaks down and we are in stall) if the pilot just lets go of the stick (Fehrm, 2018).

model (not too deep in order to spend too much time and not too shallow in order to support meaningful analysis).

A function is defined by

- A role (i.e., the task that it is capable of performing)
- A context of validity
- A set of resources enabling its realization that need to be elicited from subject-matter experts.

Function's resources are systems that have functions themselves. These functions can be cognitive or physical. Therefore, a function is a function of functions, as an agent is an agency of agents (Minsky, 1986).

Figure 5.4 presents an HCD method that enables the construction of a function model by carrying out a task analysis that enables defining a task model, which itself serves as a placeholder for the definition of a network of functions.

Once a task model is determined, everything is ready to start a functional analysis. Placeholders are there in the form of a network of roles (determined by tasks), where function's contexts and resources should be filled in. Each function is initialized in terms of role, context, and resources from knowledge acquisition from experts and experienced people (i.e., subject-matter experts). A function model is then formed, and Human-In-The-Loop Simulations could be run, activity observed, and performance measured. An activity analysis is performed and results

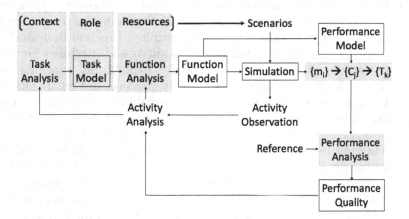

Figure 5.4 HCD using a task model, a function model by carrying out activity analysis using a simulation, and measuring performance.

injected into a new function analysis (i.e., the loop is then closed). A new function model is formed, and so on until a decision to stop is made. It is interesting to see that this approach leads to a measure of performance in a human–machine teaming environment. This method is a concrete account to incremental modeling of multi-agent systems (i.e., systems of systems) based on experience feedback using activity analysis.

Once a function model is setup, a Human-In-The-Loop Simulation can be run with respect to the execution of appropriate scenarios. There are two types of scenarios: **procedural scenarios** that provide chronological scripts of events and tasks in various context, and **declarative scenarios** that provide configurations of agents/systems and possible relationships among them.

Activity of the various agents can be observed and further analyzed. Activity analysis, combined with performance analysis, enables upgrading function analysis and resulting function model.

At this point, we should not forget that we are deploying an agile HCD approach, and therefore, there are several loops connecting the various processes (e.g., task analysis, function analysis, activity analysis, and performance analysis). Simulation provides opportunity to measure performance of human–machine cooperation (teaming). Performance assessment requires the definition of three types of entities:

- **Low-level measures** $\{m_i\}$ (e.g., workload, fatigue, attention, vigilance, engagement), also called observable variables, obtained by gathering objective data (e.g., eye gaze, electrocardiograms), subjective data (e.g., Cooper–Harper's evaluation scale, NASA TLX) and a posteriori analysis of agents' activities (e.g., auto-confrontation and commentaries on audio and video recordings).
- **Criteria** $\{C_j\}$, also called measurement models, obtained by meaningfully combining low-level measures in the form of a measurement model $C_j = f(\{m_i\})$ using physical and cognitive function analysis (Boy, 1998). Criteria are directed toward humans and machines and can be for example, affordances, usability, explicability, machine failures, delegation, flexibility (technology, practice, and organizations), maturity, supervision, authority, sustainability, tangibility, human errors, fatigue, memory, stress, vigilance, risk management, situation awareness, decision-making, and potential complacency.
- **Teaming performance metrics** $\{T_k\}$, also called high-level metrics or metrics for short, where $T_k = g(\{C_j\})$ is a meaningful contextual combination of relevant criteria. For example, these metrics can be team coherence and effectivity, trust, communication, cooperation, and coordination. These metrics need to be qualified to guaranty their consistency and operational applicability.

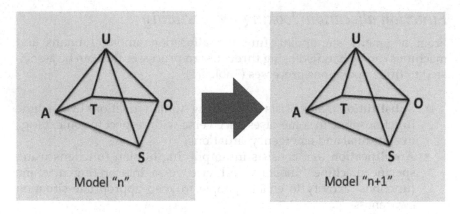

Figure 5.5 Evolution from design model n to design model n + 1 with respect to the AUTOS pyramid.

In addition, a performance model constructed from the function model provides low-level measures $\{m_i\}$, meaningful criteria $\{C_j\}$, teaming performance metrics $\{T_k\}$, and their articulations (Figure 5.5). Simulations provide quantitative and qualitative values of $\{m_i\}$, $\{C_j\}$, and $\{T_k\}$, which enable to proceed with a performance analysis. Performance analysis is carried out using a reference (i.e., a baseline that is used for comparison purposes) and $\{m_i\}$, $\{C_j\}$, and $\{T_k\}$ values to generate a performance quality model that will be further considered in a recurring activity analysis.

For example, considering the evolution of increasingly autonomous vehicles, changing from an automation level (Level n) to another (Level n + 1) requires performing task/function/performance analyses at both levels. Level n serves as the reference, and performance at Level n + 1 is compared to the reference to infer Level n + 1 performance quality. Several iterations can be implemented to refine task/function/performance models and, in the end, performance quality. Design history evolution using the AUTOS pyramid can be visualized (Figure 5.5).

Visualization of AUTOS pyramid evolution, associated with design cards, provides insightful and tangible appreciation of design and development completeness in terms of metrics being used. We are talking about design process tangibility here.

Summarizing, function analysis is crucial and should consider results from task analysis, activity analysis, and performance analysis, which contribute to the construction of a function model useful for putting together Human-In-The-Loop Simulation, which in turn supports activity analysis and performance monitoring.

Function allocation: looking for flexibility

From a general standpoint, function allocation among humans and machines can be handled using three design processes that can be associated to three operations processes (Table 5.1):

- **Substitution** that consists in replacing human functions by machine functions and, in some cases, vice versa with respect to context (e.g., in abnormal and emergency situations)
- **Amplification** that consists in amplifying human functions using specific machine functions and vice versa interpreting machine functions' activity to enable people to keep appropriate situation awareness
- **Speculation** that consists in inventing machine functions that enable people to do what they are capable of doing and discovering emerging human functions that are mandatory to deal with highly automated and autonomous systems.

Autopilots have substitution functions (e.g., they are able to follow a heading, a speed, and altitude, and so on). Pilots can perform these functions "manually." Rigid automation takes care of these kinds of functions in predetermined contexts. Therefore, substitution is directly associated with **automation** at operations time.

Context-aware search algorithms have amplification functions (e.g., they enable you to find appropriate information by typing a few keywords in context, where context is incrementally learned by the machine from previous interactions). These functions amplify your memory since they not only consider your interactions but also interactions from other people "who are like you." Therefore, amplification is directly associated with **interaction** at operations time.

Compared to birds, people are handicapped; they cannot fly! Engineers speculated aircraft functions that enable people to fly. These functions are based on physical phenomena, such as thrust and lift (Figure 5.6), that were modeled in the form of equations, which in turn were used to design

Table 5.1 Design and operations processes for human–
machine function allocation

Design process	Operations process
Substitution	Automation
Amplification	Interaction
Speculation	Augmentation

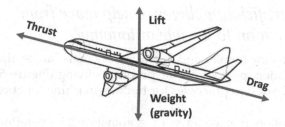

Figure 5.6 Four forces of flight (thrust opposed to drag, and lift opposed to weight).

appropriate machine structures and functions. Aircraft are prostheses that enable people to fly. In this sense, aircraft augment people's capabilities. Therefore, speculation is directly associated with **augmentation** at operations time.

It is interesting to better understand the evolution of communities from HFE to HCI to Human–Systems Integration (HSI) in the light of the triptych [automation, interaction, augmentation]. HFE was born from problems that needed to be solved in industry during the 20th century where machines were essentially mechanical and where these machines got incrementally automated. Automation became a real issue from an HFE perspective. The substitution of a human function by a machine function created the emergence of new human functions that were not necessarily anticipated. HFE specialists had to help fixing these issues. Automation turned out to introduce rigidity when things started to go wrong.

In the beginning of the 1980s, computers became to be extensively used and created new issues related to HCI. A new community was created to study these issues. People were able to do more things using computers because software amplified their capabilities. However, these new machine functions did not remove rigidity in operations. Indeed, machine functions, even if they were highly interactive, were designed in specific contexts. Outside of these contexts, these functions are not operative and may lead to major problems.

Aircraft are good examples of augmentation of people's capabilities. However, solutions that have been found so far, such as fixed wing aircraft, provide rigidity compared to what a flock of birds can do. Birds are far more autonomous and flexible than aircraft. Consequently, a new round of speculation should be carried out! More specifically, we should move from traditional single-agent to multi-agent approaches to function allocation.

How can artificial intelligence help move from rigid automation to flexible autonomy?

We have seen how human and machine functions can be allocated either using automation, procedures, or problem solving (Figure 5.2). Table 5.1 provides three design processes for human–machine function allocation:

- Substitution that deals with automation at operations time and works perfectly as long as operations stay within the context of auto- mation validity
- Amplification that deals with well-bounded interaction defined by well-defined procedures within the context of amplification/ procedures validity
- Speculation that deals with augmentation of human capabilities in order to solve problems not covered by the first two processes.

We have seen that flexible autonomy is mostly needed in abnormal and emergency situations, that is, in problem solving where people have to speculate appropriate solutions. This is precisely where AI could be effi- cient and effective by supplying tools that augment people's capabilities in problem solving.

At this point, what topics do we have within the AI community? The 2020 AAAI[8] conference[9] proposes the following traditional topics: search, planning, knowledge representation, reasoning, natural language pro- cessing, robotics and perception, multi-agent systems, statistical learning, and deep learning (DL).

As already discussed, flexible autonomy requires integration of experience and expertise, which is often available in the form of cases solved in the past and that can be reusable in similar situations. AI extensively developed knowledge-based systems (KBSs) and case-based reasoning (CBR), which can be very effective when associated with supervised machine learning (ML). Handling cases requires appropri- ate situation awareness, and therefore, AI can supply approaches such as intelligent visualization, which involves DL. Figure 5.7 presents a work- flow that integrates these AI techniques from the real world to databases and back to the real world, using a digital twin. Note that we deliberately choose a multi-agent framework that is well suited for carrying out func- tion allocation.

Developed during the eighties, **Knowledge-Based Systems** (KBS) are also known as expert systems or rule-based systems. Lots of work has been done in the field of knowledge acquisition for KBS (Gaines &

[8] Association for the Advancement of Artificial Intelligence.
[9] https://aaai.org/Conferences/AAAI-20/aaai20call/.

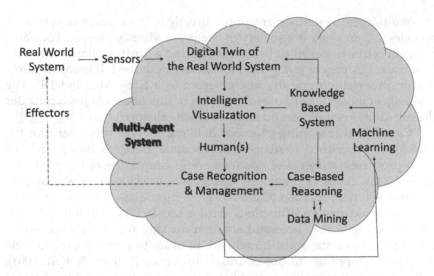

Figure 5.7 Information workflow integrating AI techniques for various types of processing, such as experience feedback management and integration, decision-making, diagnostic, and repair.

Boose, 1988; Gaines, 2013). This domain declined over the years since the beginning of the 1990s. KBS should be revived because they can support some kinds of automation of expertise and experience and, more specifically, experience feedback and integration into appropriate integrated models (e.g., in current digital twin developments).

Machine learning (ML) and even more importantly **deep learning** (DL) have become very dominant during the last few years. Obviously, ML can be very interesting within our flexible autonomy endeavor. Indeed, flexible autonomy should be based on experience, and experience is acquired incrementally through a large number of try-and-error activities. As a matter of fact, positive experience is as important as negative experience. If incidents and accidents are very well documented and can serve as useful data for learning, we should better focus on positive experience (i.e., things that went well). ML algorithms are developed to make sense of large amounts of data (i.e., the now famous "big data"). They enable to elicit patterns, organize information, detect anomalies and relationships, as well as make projections. These algorithms will enable to fine-tune the way increasingly autonomous systems will perform more safely, efficiently, and comfortably. They can contribute to improve task execution precision. DL enables the management of higher abstraction levels than ML (e.g., image and speech recognition). DL is based on many kinds of multilayer neural network technology. It enables content generation and improves existing contents, such as automated coloring of black and white images.

Multi-agent systems correspond directly to the concept of system of systems (i.e., an agent is a society of agents, in Minsky's sense). Formally, an agent can be represented by a structure and a function, hardware and/or software. An agent is a system as already described. It involves information processing (typically implemented as a KBS), ML, and DL. The multi-agent model is very useful in terms of functions when we consider the substitution-amplification-augmentation categorization.

Case-based reasoning (Aamodt & Plaza, 1994) is very useful in the context of experience feedback management. CBR[10] is based on four main processes: retrieval of one (or several) similar case(s) similar to a current case, reuse of previously used cases to elaborate a working-in-progress solution, revision of the working-in-progress solution by modifying its structures and functions until a satisfactory solution is found, and recording of the successful solution for later reuse. CBR should be develop within a statistical framework in order to perform probabilistic inference as opposed to deterministic inference (Wilson & Keil, 2001). CBR can be managed using plausible reasoning (i.e., plausibility of a case can be incrementally adjusted with respect to more evidence acquired from the real world). Jaynes showed how plausible reasoning can be based on Bayesian inference (Jaynes, 1988). Bayesian inference is a mathematical statistics technique based on Bayes' theorem (Efron, 2013), which says that the probability of A knowing B is equal to the probability of B knowing A multiplied by the probability of A divided by the probability of B. In practice, Bayesian inference consists in modifying the probability of a hypothesis when more information is available. At the same time, Bayesian classification (also called probabilistic learning) can be used to get more generic cases and structure them into belief networks (Cooper & Herskovits, 1992).

Intelligent visualization (Kruchten, 2018; Garcia Belmonte, 2016; St. Amant et al., 2001) is a growing field of investigation that attempts to develop visualization methods and tools that enable complex data to be better understood by people. In other words, it **helps people to be more familiar with complex systems** when they are appropriately visualized. Remember the adage, "A picture is worth a thousand words." In addition to static pictures, dynamic animations, simulations, and movies can be displayed to improve understanding of complex data and concepts. AI can support intelligent visualization as recent developments show in visual analytics (Keim et al., 2008). More specifically, it is highly suggested that human visual exploration should be mixed with data mining for the creation or management of existing knowledge, which in turn can be used to select the right data (Figure 5.8).

---

[10] Case-based reasoning is rooted in Roger Schank's research work on human dynamic memory (Schank, 1982).

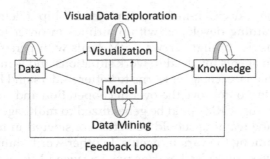

Figure 5.8 Mixed human (visual) and machine data analysis (leading to HSI). (Adapted from Keim et al., 2008).

Multi-agent co-adaptive systems: various approaches

In any case, automation (software) should be reliable at any time in order to support safe, efficient, and comfortable work. There are many ways to test software reliability (Lyu, 1995; Rook, 1990). In this book, what we try to promote is not only system reliability but also HSI reliability. We know that there is a **co-adaptation of people and machines** (via designers and engineers, as well as trainers and accumulated experience). Human operators may accept some unreliable situations where the machine may fail as long as safety, efficiency, and comfort costs are not too high (i.e., acceptable degraded modes of operations). However, when these costs become too high for them, the machine is just rejected. Again, this states the problem of product maturity (Boy, 2005); the conventional capacity maturity model for software development (Paulk et al., 1995), systematically used in most industries, does not guarantee product maturity but manufacturing process maturity. Product maturity requires continuous investment of end users in design and development processes. At the very beginning, they must be involved with domain specialists to set up high-level requirements right; this is an important role of participatory design. During the design and development phase, formative evaluations should be performed involving appropriate potential end users in order to incrementally "invent" and discover the most appropriate future use of the product in an agile way.

For the last three decades, **crew resource management** (CRM) was strongly developed in aviation, motivated by social microworld of aircraft cockpits where pilots need to know more about their personality, cooperate and coordinate to fly safely and efficiently. CRM started during a workshop on *resource management on the flight deck* sponsored by NASA in 1979 (Cooper, White & Lauber, 1980). At that time, the motivation was the correlation between air crashes and human errors as failures of interpersonal communications, decision-making, and leadership (Helmreich et al., 1999). CRM training developed within airlines in order to change attitudes and behavior of flight crews. CRM deals with personalities of the various human agents involved in work situations and is mainly focused on teaching, that is, each agent learns to better understand his or her personality in order to improve the overall cooperation and coordination of the working group. CRM could be generalized to multi-agent HSI, where human–machine teaming should be better considered in terms of inter-subjectivity[11] among the various agents. In other words, human–machine teaming involves supervised and/or unsupervised ML, as well as training of people involved in order to co-adapt people and increasingly autonomous systems.

Another field that grew up in HCI during the same period is **computer-supported cooperative work** (CSCW). The term CSCW was coined in 1984 by Paul Cashman and Irene Grief. CSCW is a multidisciplinary approach focused on how people work and how information technology could support them (Grudin, 1994). CSCW typically promotes low-fidelity prototypes and Wizard of Oz studies, always attempts to provide clear answers on which interaction techniques are best among a set, generates enough qualitative user feedback to foster formative evaluation during the design and development process, and is applicable for casual social environments (Mayer et al., 2018). Therefore, CSCW background and contributions are very interesting to support the design of systems of systems, considered as teams of teams.

In parallel with these two fields of research (i.e., CRM and CSCW), two others were developed: human reliability and distributed cognition. **Human reliability** (Reason, 1990) led to a very interesting distinction between two approaches of human reliability whether the focus is on the person or the system. Each approach induces a quite different philosophy of error management from the other. Reason stated that we cannot change human limitations and capabilities but can change task execution conditions. Therefore, these conditions, which can be viewed as technological and organizational constraints, should be clearly identified in order to create defensive barriers against the progression of an unsafe act (e.g., Reason's Swiss Cheese Model, 1997).

Distributed cognition (Hutchins, 1995) was first developed to consider sharing of meaningful concepts among various agents. Extending the phenomenological school of thought, agents are considered as subjects and not objects. They have different subjectivity, and therefore, they need to adapt among each other in order to develop a reasonable level of empathy, consensus, and commonsense sharing; this is what intersubjectivity is about. This line of research cannot avoid considering intercultural specificities and

---

[11] "The sharing of subjective states by two or more individuals" (Scheff, 2006).

differences. It is not surprising that most leaders of such a field come from anthropology and ethnology. Obviously, the best way to better understand interactions between cognitive agents is to be integrated in the community of these agents, as anthropologists and ethnologists do. In the framework of human–machine systems, we extend the concept of distributed cognition to humans and machines. The extension of the intersubjectivity concept to humans and machines requires that we consider end users and designers in a participatory way. To summarize, human factors mainly deal with *user's knowledge, skills,* and *expertise* on the new artifact and its integration.

By definition, increasingly autonomous systems are incrementally modified by adding emerging functions and potential structures. The more a system is automated, the more new functions and structures emerge and need to be identified, understood, and made transparent to both design and operations communities. These modifications require an evaluation and validation plan that should consider maturity of involved TOP's activity. In other words, we need to manage these maturities and change management that goes with this constant evolution. Such maturity can be measured in terms of reliability, robustness, persistence, predictability, and ease of anticipating and understanding mode transitions (i.e., nonlinearities).

Requirements for sociotechnical systems flexibility

When we think sociotechnical systems in terms of multi-agent systems or systems of systems, we necessarily have to consider the required level of *coupling* between the various purposeful agents/systems that need to handle generated sociotechnical activities and possible changes. Organization complexity is linked to social cognition, agent–network complexity, and more generally, multi-agent management issues. **Sociotechnical systems' flexibility** is a matter of communication, cooperation, and coordination among the various agents involved. Let's propose four principles for sociotechnical systems' flexibility in terms of agent/system activity:

- Awareness (i.e., know what the other agents are doing now and for how long)
- History (i.e., what all agents have done so far)
- Rationale (i.e., why agents are doing what they do)
- Intention (i.e., what other agents are going to do next and when).

These principles can be thought in terms of system function's role. Indeed, role analysis (i.e., task analysis) directly leads to activity/job analysis, which is very useful to define qualifications of personnel and capability of technology required to perform required tasks within an organization or environment.

What we can learn from crisis management

Automation and HCI turn out to be rigid and sometimes counterproductive in abnormal and emergency situations, or unexpected and unknown situation. Flexible technology, which I call **FlexTech**, should be designed to support such situations. Therefore, FlexTech should support problem solving.

Considering the TOP model, FlexTech should be designed along with these three dimensions: technology, organizations and people. In other words, FlexTech should support people and organizations, as well as insure continuity with what automation and HCI supports. In addition, people and organizations should understand FlexTech to solve problems in such situations. On the one hand, people should have appropriate training, skills and knowledge on FlexTech support, and on the other hand, organizations should be appropriate culture on FlexTech. Consequently, developing FlexTech is a threefold endeavor.

Let's take the example of crisis management after an earthquake and, more specifically, Fukushima Daiichi nuclear disaster that occurred in 2011 in Japan.[12] We will not describe the event again here, many other authors did it before (Travadel & Guarnieri, 2015). The nuclear operator company, Tepco Group, and the Japanese government have been ineffective during the crisis management that followed the earthquake and tsunami. Tepco was unable to prevent a nuclear meltdown and subsequent explosions. The entire situation was exceptional. This disaster suggests the following questions. What skills and techniques are relevant to handle such situations? What kind of resilience mechanisms should be put in place? How should TOP jobs/activities be redesigned in such dangerous and very rare situations? How do we deal with the unknown?

Let's try to partially answer the questions above. **Intelligent visualization** is certainly a way to go to increase situation awareness of people involved in crisis management. Of course, not all possible data are available, but data that are available should be visible in an integrated fashion, which provides meaning. HSI should be developed at the highest level, down to the lowest level, where human information processing and machine information processing should be supported in a collaborative way (Figure 5.8). The system of systems approach should be developed, where simulation data should be combined with real-world data to make projections and support decision-making. Intelligent visualization that insures up-to-date and cross-checked information is technology that provide flexibility because decision makers can build reasoning on data they can trust. This is a way to consider FlexTech (i.e., support to problem solving and, in this case, in highly pressurized situations).

---

[12] Investigation Committee on the Accident at the Fukushima Nuclear Power Stations of Tokyo Electric Power Company. *Final Report* (2012).

What we can learn from people involved in life-critical systems

Human operators involved in life-critical systems require very important *skills* such as creativity, familiarity, availability, adaptability (or flexibility), dependability, and boldness (Boy, 2013b). Indeed, any actor who needs to face unexpected situations is required to be

- Creative and foreseeing possible futures; for example, when Captain Sullenberger decided to land his Airbus 320 on the Hudson River on January 15, 2009, he was creative and, for sure, investigated all other possibilities before taking the risk (NTSB, 2010)
- Familiar with the environment where they work; for example, flying skills in various atmospheric situations and aircraft configurations
- Familiar with the various tasks that they have to perform; for example, normal and abnormal tasks experienced in a flying simulator
- Familiar with personal capabilities and limitations; for example, reduced perception of night situations while driving or working memory cognitive limitations
- Familiar with organizational constraints and possibilities; for example, responsibility and accountability related to a job in an organization
- Familiar with technological constraints and possibilities; for example, automation limitations and advantages in a large variety of situations
- Available anytime anywhere during duty time; for example, management of complacency in case of routine activities and maintenance of proactive behavior
- Adaptable (or flexible) to any operational situation; for example, facing an unexpected event such as wind shear, pilots will fit their behavior with respect to changes in their environment; they know the various contextual responses to wind shear[13]
- Dependable in life-critical situations; for example, a mountain guide is typically trustworthy in dangerous situation with his or her clients
- Bold in risk taking; for example, facing an unexpected life-critical situation, a human operator should have the courage to take an appropriate action that may put his or her life in danger.

Among the many things that we learned from **risk taking** and complexity management (Boy & Brachet, 2010) are: preparation is key, becoming familiar with complexity that needs to handled is also key, routine is a killer, and so on. How can these findings help in designing and developing FlexTech? First, flexible technology should be equipped with ML mechanisms that incrementally make them more familiar with their

---

[13] www.skybrary.aero/index.php/Low_Level_Wind_Shear.

environment. Second, flexible technology should constantly inspect and provide situation awareness to people and systems involved in the management of the situation. Therefore, FlexTech is about constant smart monitoring and learning. Note that learning could be done automatically or supervised by appropriate people. Research should be further developed on this topic to better determine what technology (as a resource) should or could learn by itself and not. In the majority of cases, we can expect that humans and machines will work together to learn, keeping the validation process in the hands of experts and operational people.

Flexible technology should be easily usable and modifiable

Flexible technology cannot be developed and certified in one shot. It has to be incrementally constructed using a trial-and-error approach. **Testing is a major process** in making flexible technological system a useful and usable resource. Usefulness can be described in terms of result effectivity and satisfaction of expectations. Flexible technology will be useful when it will support successful endeavors, especially in abnormal and emergency situations as well as unknown and rare situations.

Usability is a matter of ease of learning by people who will use the system being considered; ease of retention of what has been learned; efficiency of the system; ease to recover from human errors and machine failure; and satisfaction in terms of pleasure, aesthetics, and "cool spirit."

In addition, even if it has to be delivered at some point, we can say that flexible technology is never finished and, therefore, needs to be easily modifiable. This modifiability aspect can be seen short term and long term. Short-term modifiability concerns cases where a solution proposed by the CBR processor, for example, should be adapted to the current situation. We will talk about short-term flexibility. Long-term modifiability concerns modification of case structures and functions stored in KBS's long-term memory (Figure 5.8).

Community of autonomous systems

As already said, the more agents or systems interact among each other, the more they need coordination rules. In some situations, they need to understand what the others are doing. When two agents disagree on something (e.g., a driver and a highly automated system disagree on passing a car), the human can explain the reason why he or she wants to act, as the machine will not be able to explain anything because nothing is programmed into it to this end. Therefore, human–machine teaming requires cooperation in the sense of sharing action rationale among agents involved. Consequently, explanation capabilities should be developed in machine autonomous systems. Autonomy goes with shared situation awareness.

An unfortunately good example of disagreement between human and machine is what happened in China Airlines Flight 140, A300B4-622R, B1816. After an uneventfully flight from Taipei, Taiwan, the aircrew started the approach to land on Nagoya airport in Japan on April 26, 1994. The pilot flying (PF) was the first officer who inadvertently activated the go-around switches on the throttles. Consequently, the auto-throttle go-around mode was activated, which caused a thrust increase and a climb above the glide path. PF attempted to return to the glide path using forward yoke. This was the disagreement between the human (i.e., the PF) who attempted to descent and the machine (i.e., the autopilot system) that attempted to climb. Of course, the PF did not realize that he made a mistake by activating the go-around switches, but the aircraft did not understand either and did what it was ordered to do. When the aircrew realized that they could not overcome the greater aerodynamic force of the stabilizer, the crew initiated a go around (now intentionally). Unfortunately, this new application of thrust caused a greater pitch-up and an airspeed decrease... aircraft stalled and crashed.

Now that we speak more about autonomous systems, it is time to talk about human–machine cooperation. Context should be considered more than before. For example, in the case of Nagoya accident, knowledge of the flight phase would have been crucial. Activation of the go-around switches inadvertently around 1,000 feet is a serious concern, especially when the PF tries to return on the glide slope. Detection of this contradiction should be possible, and a message (auditory) could be generated saying, "Go Around started, impossible to descent any longer!" This is an example of explanation given by the machine. A more affordable message would be to have a strong flashing light on the throttle where the go-around switches are. Emission of both messages at the same time would be ideal because redundancy is always good in case of emergency.

Human–machine co-adaptation is very difficult because these two types of agents (i.e., humans and machines – natural and artificial systems) are not made from the same premises. On one side, human behavior is almost impossible to quantify and cognitive functions are difficult to model. They are either very low level (i.e., the neuronal level) and difficult to scale up to consciousness, for example, or high level (i.e., conceptual models of human cognition) and both subjective and too macroscopic to handle useful details. On the other side, automated machines have become more complex and reliable over the years, designed using very strict rules and STEM[14] principles. For sure, life, human, and social sciences should be cross-fertilized with STEM. This is one of HSI goals.

---

[14] Science, Technology, Engineering, and Mathematics.

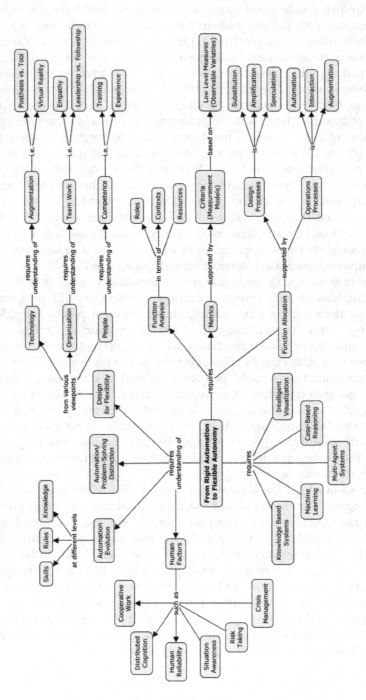

Figure 5.9 From rigid automation to flexible autonomy.

Summary

Moving from rigid automation to flexible autonomy requires understanding (Figure 5.9): automation evolution in terms of skills, rules, and knowledge; the distinction between automation and problem solving; design for flexibility in terms of TOP; and human factors such as cooperative work, distributed cognition, human reliability, situation awareness, risk taking, and crisis management. It also requires carrying out function analyses and function allocation, as well as defining metrics based on criteria and low-level measures. Function allocation is typically supported by design processes (i.e., substitution, amplification, and speculation) and operations processes (i.e., automation, interaction, and augmentation). Finally, moving from rigid automation to flexible autonomy requires implementation of KBSs, ML, multi-agent systems, CBR, and intelligent visualization.

chapter six

Orchestrating Human–Systems Integration

> We believe that the essence of a system is
> 'togetherness', the drawing together of various parts
> and the relationships they form in order to produce
> a new whole…
>
> **John Boardman and Brian Sauser**
> *System Thinkers*

Should Human–Systems Integration (HSI) be a discipline? Human factors and ergonomics (HFE) specialists say that they are complementary to engineers and, therefore, do not advise to promote HSI as a discipline. Engineers promote systems engineering from a technological point of view and, for a long time, have usually assumed that HFE can be done once product development is done. With tangible virtual human-centered design (HCD), HFE, human–computer interaction (HCI), and systems engineering naturally leads to HSI, as a process and a product. From a systems engineering perspective, Folds's definition suits this book purpose: "HSI is a set of systems engineering processes that ensures all human-related technical issues are properly identified and addressed during system planning, design, development, and evaluation" (Folds, 2015).

Therefore, as a radical statement, HSI cannot be reduced to an element of systems engineering, it should encapsulate systems engineering (i.e., we cannot do systems engineering without concurrently considering technology, organizations, and people at the heart of all sociotechnical analyses and product life cycles).

HSI is crucial because it promotes **integration** from the very beginning of the life cycle of a system, and Virtual HCD seriously considers the human element in the design of the complex system. We then need to train people who will be effective to coordinate collaborative work and produce a real integrated product. In other words, taking music

125

as a metaphor, we need both composers (i.e., people who are making scores – procedures and practices) and conductors (i.e., people who are in charge of coordinating the organization of agents performing). Pew and Mavor already recommended that "developing HSI as a discipline and preparing HSI specialists to be system development managers" (Pew & Mavor, 2007).

Integrated design associated with participatory design

For a long time, I try to put together HSI and HCD, where the common denominator is the "human." However, concepts of "integration" and "design" are as essential to the approach developed in this book. Two issues that we often observe in industry are: integration of the parts of a system is often done too late; and requirements of a complex system are often defined too fast and lead to heavy repairs of design flaws later in the development process when actual integration is done. In other words, integration should be anticipated during the early stages of engineering design. We call this process "integrated design."

Integrated design is not a new kind of approach and process (Tichkiewitch & Brissaud, 2013). It is holistic (i.e., it brings together several disciplines that are often considered apart from each other). For example, mechanical engineering, computer science, human and social sciences, and architecture can be put together and coordinated at design time to make an integrated design concept and requirements for later development. Also, integrated design can be considered as design of the development process and its final solution.

Around the 1960s and 1970s, **participatory design** was heavily practiced in Scandinavian countries and, more recently, reinvented in the HCI community (Gregory, 2003). Various participatory design methods have been developed (Schuler & Namioka, 1993; Greenbaum & Kyng, 1991). For example, they are focused on ethnographic field research, cooperative prototyping, and design by doing. Both integrated design and participatory design are called co-design, and even if they are not rooted in the same background – the former being rooted in loose coupling of requirements produced by various professions around the same product, and the latter in cooperative work including unions and social entities – I propose to merge them in this book. Participatory design increases cooperative work at design and development time, and integrated design reduces dependencies at the solution level.

It is interesting to notice that the combination of integrated design and participatory design could name **"charette design,"** as a very intense design activity, which Émile Zola described in *L'Œuvre* (The Master Piece) as the "charette night" to denote the finishing touches of a design process produced by fine arts students at the *Ecole des Beaux Arts* of Paris who were putting their designs on a cart[1] rolling among them, for review purposes (Zola, 1886).

From the Old Army model to the Orchestra model

The metaphor of the orchestra has been described already in this book and elsewhere (Boy, 1991b, 2013a). However, it is important to understand why it is so important today in industry, and in our society in general. This is the reason why the current shift from the Old Army model (OAM) to the Orchestra model should be explained at this point.

Most industrial organizations work according to the OAM (i.e., a general at the top, then officers and soldiers at the bottom). OAM information flow is vertical and mostly top-down. Soldiers do not or barely exchange information horizontally. They are executants of low-level tasks. Decisions are made at the top.

For the last few decades, horizontal information flow exists supported by information technology (e.g., phones, emails, Internet). This kind of information technology-based horizontalization happens anywhere and, more specifically, in industrial organizations. Consequently, the very concept of OAM has to evolve. In addition, the soldiers of the past are now highly specialized. They have become experts in a given field of practice. They even often belong to communities of practice. The overall organization concept shifts toward the Orchestra model.

For an orchestra to be effective and enable playing a symphony, all musicians are required to have a common frame of reference, which is music theory (e.g., they know how to read and understand the meaning of scores). Who is writing scores? This is the dedicated job of composers. A composer of a symphony needs to articulate various scores among each other. He or she has to coordinate tasks of musicians of the orchestra. More generally, this is task coordination for an organization to be able to perform some kind of product. At performance time, the orchestra requires a conductor to synchronize musicians' activities. In other words, the conductor coordinates a system of systems (i.e., an orchestra of musicians). Musicians themselves are autonomous systems or agents capable of playing their part perfectly but need their activity to be coordinated with the other musicians. Again, activity may be very different from the task. In addition, musicians also need to cooperate among each other to insure a reasonable amount of stability and resilience in case of a musical

[1] A "charette" is a cart in French.

mistake by one of several musicians – "One for all! All for one!"[2] Another part of the orchestra metaphor is the audience – music stakeholders (i.e., composers, conductors, and musicians) produce pieces of music for potential listeners! More generally, designers, crafters, and engineers generate products for targeted people (i.e., we often call them users or human operators). The audience brings issues such as acceptability, usability, and usefulness. It can be made of anybody (i.e., a general public audience) or experts.

In the same way, we have several types of orchestras (e.g., symphonic, jazz), and current industrial organizations may take several forms (e.g., highly structured and large, loosely structured and small). Structured and large organizations are typically based on proceduralized functions (e.g., symphonic orchestra). Loosely structured and small organizations are typically based on problem-solving functions (e.g., jazz band). In all cases, we are facing agencies of agents (i.e., systems of systems or teams of teams), and the more agents are autonomous, the more the agency should be coordinated.

Developing an ontology of a system of systems

First thing first! It is important to develop an ontology[3] of the domain at stake. Why? Participatory design requires everyone to understand each other. Therefore, they need to speak the same language. Defining a common language is difficult because we need to gather the various concepts defining the domain. This language should provide common understanding of information to be shared by the various human and machine systems (or agents). It should be evolutionary because it is impossible to get a definitive ontology but a stabilized explicit ontology. This language should support analysis, design, and evaluation of systems of systems.

What is an ontology in HCD of complex systems? The atomic element of an ontology is the "concept." Concepts can be incrementally elicited in many ways: (1) from stories told by subject-matter experts or analysis of questionnaires, (2) from observations of people and systems involved in the domain at stake, and (3) from brainstorming sessions

---

[2] "Un pour tous, tous pour un" in French is the famous phrase of *The Three Musketeers*, a novel of Alexandre Dumas (www.crdp-strasbourg.fr/je_lis_libre/livres/Dumas_ LesTroisMousquetaires.pdf, retrieved on 21 June 2019) and the unofficial motto of Switzerland.

[3] In artificial intelligence, an ontology is defined by the set of concepts of a domain of interest, their properties, and the relations (links) among the various concepts (Musen, 1992; Gruber, 1993a), in the same way a terminology is defined by a set of terms (i.e., like a dictionary). Cognitive science defines a concept by associating a term to it and its meaning (i.e., context that removes possible ambiguities). More generally, "ontology is the philosophical discipline that tries to find out what there is: what entities make up reality, what is the stuff the world is made from?" (Hofweber, 2005).

associating domain stakeholders and HCD teams. This process is typically called **knowledge elicitation** or **knowledge acquisition** from experts (Gaines, 2013).

Gruber proposed five design principles and criteria for ontologies whose purpose is knowledge sharing and interoperation among programs based on a shared conceptualization (Gruber, 1993b). Let's propose an adaptation of these five principles:

1. **Clarity** expressed in the form of effective and objective terms and relations among them (i.e., keep intended meaning). All definitions should be documented with natural language.
2. **Coherence** between formally defined concepts and links among concepts and their definitions (i.e., what you see is what you get – inferences should keep consistency between formal concepts and definitions expressed in natural language).
3. **Extendibility** of the ontology in such a way that enables an HCD team to extend and/or specialize it monotonically (i.e., new terms and links could be added easily).
4. **Minimal encoding bias** that guaranties that the ontology should not be [too much] dependent on the way it is coded (i.e., conceptualization should be specified at the knowledge level without depending on a particular symbol-level encoding).
5. **Minimal ontological commitment** to support the intended knowledge sharing activities (i.e., an ontology should make as few claims as possible about the world being modeled – it should be easy to specialize and instantiate the ontology when needed).

Knowledge acquisition from experts is a matter of **knowledge design**. Design is a matter of making compromises or trade-offs. Ontology design is not different. We will then have to make compromises among the abovementioned principles. "For example, clarity criterion talks about definitions of terms, whereas ontological commitment is about the conceptualization being described. Having decided that a distinction is worth making, one should give the lightest possible definition of it" (Gruber, 1993b).

Coherence can be further expanded into lexical, syntactic, semantic, and pragmatic coherence. **Lexical coherence** is about terms that are used to denote concepts within the developed ontology. When a concept is denoted by two terms, then there is no lexical coherence. We often say that it is ambiguous. In natural language, we try to vary the use of words in order to avoid boring repetitions. However, in technical writing, it is better to use the same words to denote the same concepts in order to avoid confusion. When terms are formed of multiple words, aliases are fine and often used; this is common practice. **Syntactic coherence** is about sequences of terms. For example, on a desktop user interface, it is typical to see on the top menu bar, "File – Edit – View …", under "File", we have "New – Open – Open Recent …" If a new application is created and the sequence of options is "Open – New – Open Recent …," then there is no syntactic coherence. **Semantic incoherence** is about concepts that are denoted by the same term. For example, the concept of "table" can denote a "table of contents" in a book or a physical table on which you can eat. When this is the case, you need to add context around the term to eliminate the potential confusion. For example, if you say "table of contents" or "the table in the book," we are likely to understand that it is a figurative table and not a physical table. If you say "people seating around the table," we are likely to understand that it is a physical table. **Pragmatic incoherence** is about a concept that has different meanings in different cultures or ethnical groups.

Extendibility is related to coherence and clarity. Indeed, by adding, removing, or reformulating a concept in an ontology, we need to make sure that consistency is insured as well as the various distinctions are meaningful and sustainable.

For a system-of-systems ontology, it is typically easy to define an initial structure of structures, as well as allocating functions of functions onto it (i.e., a deliberate function allocation). However, it is more difficult to find out new emerging structures and functions that should be incrementally added, removed, modified, split, merged among each other, and so on into the ontology. This is why the minimal encoding bias principle should be carefully respected.

Associating divergent and convergent thinking

Considering the ontology perspective, we now need to provide the process for incremental derivation of relevant concepts and interrelations among them. As already said, HSI concepts will be defined in the form of systems, themselves defined as structures and functions. **Systemic ontology** is typically generated in the form of abstractions to incrementally evolve toward more tangible interrelated concepts.

The first step consists in implementing a **divergent thinking** (DiTh) process in the form of brainstorming, storytelling, storyboarding, design thinking, creativity sessions, and so on. DiTh is a method that supports the spontaneous and nonlinear generation of creative ideas by exploring as many possible solutions as possible. DiTh is highly facilitated and productive when participants have personality traits such as nonconformity, curiosity, willingness to take risks, and persistence.[4] DiTh is an out-of-box approach.

[4] https://en.wikipedia.org/wiki/Divergent_thinking.

The second step consists in implementing a **convergent thinking** (CoTh) process in the form of sessions of rationalization, evaluation, and synthesis, considering the five principles: clarity, coherence, extendibility, minimal encoding bias, and minimal ontological commitment. CoTh supports crisp, simple, prioritized, and well-established solutions that should be unambiguous. CoTh is usually strongly based on standards and probabilistic voting mechanisms toward finding the best solution.

DiTh and CoTh approaches are often used in concert. As many possible solutions are elicited and designed first (DiTh), and the best solution is derived second (CoTh). The following method is a good example of a good combination of DiTh and CoTh.

The Group Elicitation Method (GEM)

> The most successful designs result from a team approach where people with differing backgrounds and strengths are equally empowered to affect the final design.
>
> *(Tognazzini, 1992 page 57). We*
> *usually organize meetings!*

There are five classical types of meeting activities that need to be further emphasized: idea generation, issue discussion, negotiation, conflict resolution, and team building. It is, however, recognized that meetings are time-consuming, inefficient, and not adapted to reaching consensus. In addition, they provide poor or incomplete consideration of alternatives, unequal participation (domination of a few members), a lack of meeting memory, and a lack of satisfaction with the meeting process (Neal & Mantei, 1993).

The Group Elicitation Method (GEM) was designed and developed during the 1990s to support knowledge acquisition for knowledge-based systems (Boy, 1996, 1997a). It was an alternative to elicitation and knowledge representation from a single expert. The GEM is very useful and effective for real-world multi-expert knowledge elicitation where cross-feedback and consensus-seeking from the experts are required.

For example, several experts may work together to investigate appropriate solutions to a design or evaluation issue. A crucial problem is to derive an acceptable consensus from a group of experts who share neither the same background nor the same objectives. It is not uncommon that experts do not understand each other. The brainwriting technique was introduced more than two decades ago to facilitate the generation of ideas or viewpoints by a group of people (Warfield, 1971). Brainwriting enables a group of people to construct a written shared memory. Each participant faces a sheet of paper and reads the issue to be investigated. He/she then adds several viewpoints and puts it back on the table, where the set of papers constitutes a shared memory of the meeting. The process of choosing a piece of paper, reading, writing viewpoints, and replacing the paper on the table is continued until each person has seen and filled in all the papers. Then, each person is continually confronted with the viewpoints of the others and can react by offering a positive or negative critique and new viewpoints. Generally, a considerable number of viewpoints can be amassed with this procedure.

A typical GEM session takes a full working day (i.e., five effective hours and breaks) and consists of six phases conducted by a knowledge elicitation facilitator:

- Issue statement formulation and choice of the participants
- Viewpoints generation (brainwriting)
- Reformulation of these viewpoints into more elaborate concepts
- Generation of relations between these concepts
- Derivation of a consensus
- Critical analysis of the results.

Issue statement formulation and choice of the participants

The first crucial and difficult phase is to state the issue to be investigated by a group of participants. It often takes a few application domain experts to state this issue. The type of checklist proposed by Nielsen (1986) for semi-structured interviews is a good starting point for the formulation of issue statements. The GEM is currently implemented using the following checklist as a guideline:

- What is the goal of the system that we plan to design or evaluate?
- How is the system or its equivalent being used or activated? (current practice, observed failures, and human errors)
- How would you use or activate this system? (usages requirements)
- What do you expect will happen if the corresponding design is implemented? (e.g., in terms of productivity, aesthetics, and safety issues)
- How about doing the work this way! (naive and/or provocative suggestions)
- What constraints do you foresee? (pragmatic investigation of the work environment).

Six to ten people (typically subject-matter experts) are chosen to participate in a GEM session. The GEM experience during the last thirty years suggests that the optimal number of participants for an interesting session in a reasonable time frame is about seven.

Viewpoints generation (brainwriting)

The issue statement is then given to the participants, and they are asked to provide their opinions or viewpoints on the formulated issue. This phase is silent – participants are writing. An example of an issue statement could be the following:

- Could you write problems that you think are illustrative of air traffic management functions from the perspective of the design of self-separation system?

For the first ten minutes, each participant writes a list of viewpoints on his/her sheet. Then each participant passes his/her sheet to the person next to him/her. At this stage, each participant faces a list of viewpoints generated by someone else (Figure 6.1). There are three possibilities of action: agreement, disagreement, or new viewpoint. In the first case, the participant mentions his/her agreement and may add more comments reinforcing the original viewpoint. In the second case, each participant mentions his/her disagreement and explains why he/she does not agree with the original viewpoint. In the third case, he/she just adds

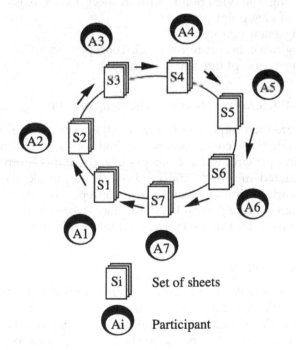

Figure 6.1 Example of a brainwriting session where sheets are turning from one participant to another.

a new viewpoint to the list. This process continues until all participants have seen all sheets. One of the main advantages of this method is that participants are not influenced by outspoken people as is often the case in conventional meetings. This phase takes usually one hour for seven participants.

Reformulation of these viewpoints into more elaborate concepts

These viewpoints are then analyzed and reformulated into a list of concepts. This phase consists in an oral meeting supervised by the knowledge elicitation facilitator, who is free to create the concepts that he/she feels are the most appropriate. Participants use their critical thinking to contest elaborated concept – they can react to and propose any change. Concepts can be refined as the reformulation process goes on. Creation and reformulation of concepts are typically performed as a collaborative process involving the participants and the facilitator. This phase may take between one and two hours. It involves four types of operations that correspond to the concept clustering mechanisms that were described and used in the COBWEB machine learning system (Fisher, 1987):

- Classifying the viewpoint with respect to an existing concept (a class of viewpoints)
- Creating a new concept
- Merging two concepts into a single concept
- Splitting a concept into several concepts.

Generation of relations between generated concepts

Participants are then requested to provide their opinions on the relative priorities among these concepts. One method consists in filling out a triangular matrix presenting the concepts in rows and columns. Basically, a score is entered into each matrix "box." For example (Figure 6.2), a participant starts at line 1, second box, if concept 1 is more/equally/less important than concept 2, then he/she writes +1/0/–1 in the corresponding box, and so on. This phase takes about thirty minutes.

Derivation of a consensus

A consensus is derived using priority matrices provided by the participants. We call global score of a relation the sum of all the scores of one relation among the participants. The global score matrix is the sum of all the matrices generated by all participants. To each global score is attached a standard deviation measuring the inter-participant consistency of the global score of the relation.

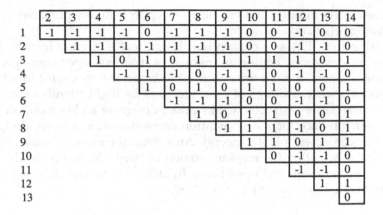

	2	3	4	5	6	7	8	9	10	11	12	13	14
1	-1	-1	-1	-1	0	-1	-1	-1	0	0	-1	0	0
2		-1	-1	-1	-1	-1	-1	-1	0	0	-1	-1	0
3			1	0	1	0	1	1	1	1	1	0	1
4				-1	1	-1	0	-1	1	0	-1	-1	0
5					1	0	1	0	1	1	0	0	1
6						-1	-1	-1	0	0	-1	-1	0
7							1	0	1	1	0	0	1
8								-1	1	1	-1	0	1
9									1	1	0	1	1
10										0	-1	-1	0
11											-1	-1	0
12												1	1
13													0

Figure 6.2 Example of a priority matrix generated by a participant.

This phase leads to the expression of a consensus among the participants (at the time they were consulted). This consensus is expressed using four types of typical parameters (normalized with respect to the number of participants and the number of generated concepts).

The mean priority (MP) of a concept corresponds to the mean of the scores assigned to a concept with respect to the other concepts for all the participants. The value range of the MP is the interval [−1, +1]. This enables the distinction of the "positive" concepts from the "negative" concepts. For instance, if all participants provide a + 1 score for a concept compared to all the other concepts, then the MP of this concept is +1.

The inter-participant consistency (C) of a concept corresponds to the mean of the standard deviations of all global scores. It is related to a consensus on the choice of subjective criteria. It conveys a degree of confidence in the MP with respect to the set of participants.

The MP deviation or stability of a concept (D) corresponds to the standard deviation of the MP with respect to the global scores of a concept. This parameter is useful to better characterize the priority concepts. The smaller D is, the more stable the group judgment is.

The global consensus (GC) expresses a global score of the group consensus on the investigated issue. This parameter is useful for comparing several groups.

Critical analysis of the results

The GEM is a semiformal method that participants tend to easily accept and even rely on to enhance cooperation. We have observed that critical analysis of the generated concept network is very constructive and reinforces the consensus. Experience shows that this phase may take between thirty minutes and one hour. A report is finally prepared by the knowledge elicitation facilitator.

Using the GEM as a conceptual design tool is a great incentive for gathering designers, engineers, managers, trainers, operations people, and maintainers. It has been observed to be extremely useful in a large variety of cases (e.g., design of user interfaces for flight attendants, design of a maintenance instructor workstation, design of a classroom, design of a user interface for a new aviation communication system, design a new nuclear power plant concept). An essential common factor among these cases was that the implementation of the GEM helped bridge the gap between design and operations. In other words, the GEM is highly recommended to support HCD and HSI.

Rehabilitating educated common sense

Orchestrating HSI is mostly a matter of educated common sense, which is simply doing things that make sense according to a well-understood purpose shared by an educated community of people (Boy, 2013). I remember how shepherds in the Pyrénées mountains of my childhood were predicting weather using proverbs and educated observation of the sky, based on a long tradition of the Arts of Memory inherited from the ancient Greeks (Yates, 1966), which consisted in associating abstract concepts together with concrete objects. This was educated common sense.

Very recently, I heard Thierry Marx, one of the best French Chefs in the world, interviewed by a journalist who asked him why his cooking was so appreciated. He simply said: "I use the RER approach!" RER means "rigor, engagement, and regularity." Indeed, delivering a great meal deserves rigor (i.e., you need to know how to make things precisely and correctly). Knowing procedures and recipes is important, but being engaged in making the meal is crucial. It is the way you cut, assemble, and finally cook that will make the difference. You need to act, taste, and finally assess if result is good or not... if it corresponds to the standard you want to reach. Regularity is matter of repeating the same kind of gestures all over again that enable you to reach the same standard in the end. This is educated common sense.

In HCD, getting good HSI results is a matter of mastering the most appropriate association of abstract concepts and concrete objects (i.e., making a relevant conceptual model of the system you want to design and use it for the development of a prototype that will need to be tested). Experience is of course crucial. Experience should be combined to creativity to make a good integration (i.e., in the same way a chef integrates various kinds of ingredients to make a great meal). From that point of view, in addition to having a vision (i.e., foresee possible future), you will need rigor, engagement, and regularity (i.e., elaborate these possible futures and test them incrementally in an agile way). This is educated common sense.

Structure–function symbiosis and minimality

It is often important to be inspired by what biology can provide. The human respiratory system is a good example of structure-function symbiosis and minimality. Human lung's function is to provide oxygen from the air to the blood and exhaust carbon dioxide from the blood to the air. Lung structure is organized into twenty-three generations of branches going from the trachea to the alveoli (at the bottom of the trachea there is a first generation of branches, one going to the right part of the lung and another to the left part; each of these branches typically divides into two sub-branches, and so on twenty-three times). A generally accepted mean number of alveoli is 480 million with a variation of 37% among people (Ochs et al., 2004). The total surface area of the cumulated membrane between alveoli and blood capillaries is about 75 square meters (i.e., a surface area as large as half of a tennis court).

Weibel's model is commonly used to represent lung geometry (Weibel, 1963). Using this model, Figure 6.3 displays cumulated diameter of the twenty-three generations of the tracheobronchial tree geometry (vertical axis, going from around 2 centimeters at the mouth to

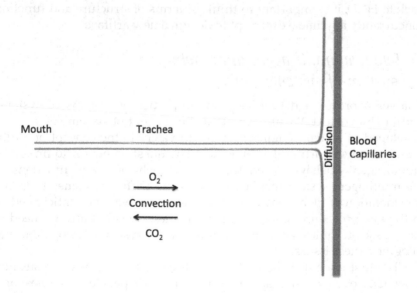

Figure 6.3 A useful model of the essential structures and functions of the human lung.

around 8.7 meters) against the linearized cumulated trajectory from the mouth to blood capillaries (horizontal axis, with a distance from the mouth to the alveoli of around 27 centimeters). It is clear that the first part of the lung structure from the mouth to around the 21st bronchial generation, fluid dynamics is mostly a convective phenomenon. Since the cumulative section is growing exponentially after that, fluid dynamics almost instantly bifurcates to a diffusive phenomenon. Consequently, the structure of the human lung enables air to transfer very quickly from the mouth to the alveoli and leave time for the diffusion process to and from the blood to take place. Of course, there are other factors that enter into play, but they are not essential for the purpose of this book. This is a great example of how nature combines structure and function to provide appropriate phenomena to happen effectively.

The human lung is a natural autonomous system that was modeled and simulated on a computer (Boy et al., 1980; Boy, 1981). We managed to anticipate new findings that were verified experimentally (e.g., the combination of mass and momentum transfer phenomena induces specific behaviors on expired gas profiles that were not explained before). Such fundamental physiological results could not be obtained so clearly without such a modeling and simulation approach integrating structures and functions.

As a general principle, structures determine physical functions and conversely. I claim that this is true for any physical architecture. Of course, several functions may take place within a single architecture. Back to HCD, it is important to think in terms of structure and function concurrently anytime we attempt to design a new artifact.

Ontological multi-agent approach: the system-of-systems concept

Chapter 4 provided definitions and properties of systems of systems within HSI context. We also saw that the concept of systems of systems developed in systems engineering is equivalent to the concept of multi-agent systems in artificial intelligence. The question now is to link individual agent's behaviors and functions to behaviors and functions of the multi-agent system they are evolving in. Difficulty comes from the consideration of global emerging behaviors that were not anticipated in individual procedures and coordination rules. Consequently, we need to have a sociotechnical approach for multi-agent system development that integrates these issues.

The first issue that should be considered is homogeneity versus heterogeneity of agents (i.e., systems). They may be people, machines, or a combination of both. They may be problem solvers, simple procedures executants, or both. They may have differing cultures (e.g., a coalition of several armies). For that matter, systemic interaction models presented in Chapter 4 should be seriously considered.

An agent is defined by a structure of structures and a function of functions. Each function is defined by its role, a context of validity, and a set of resources (that can be agents themselves). Developing a multi-agent ontology consists in incrementally defining and articulating such roles, contexts, and resources. Figure 6.4 presents a process that enables the development of such an ontology that would support the targeted multi-agent (system of systems) architecture. It starts from a purpose (i.e., a goal) that needs to be satisfied. Then, two types of scenarios are developed: declarative (i.e., describing configurations of interconnected agents) and procedural (i.e., describing chronologies and, therefore, agents functions). The former will lead to the development of a declarative ontology of the domain (e.g., Protégé can be used to support development). The latter will lead to the development of a procedural ontology of the domain (e.g., skeletal plans can be used to support development). These developments of scenarios and ontologies start by an analysis of the existing (i.e., current practice) and follows by imagining possible futures. These analyses should be done cooperatively with domain experts (i.e., experienced people). The next step is the induction of agent classes and generic relationships among them. Each time a version of the overall multi-agent ontology is achieved, the purpose should be refined. If corrections are judged to mandatory, then an iteration is made, otherwise the process is completed and can serve to support the multi-agent system architecture.

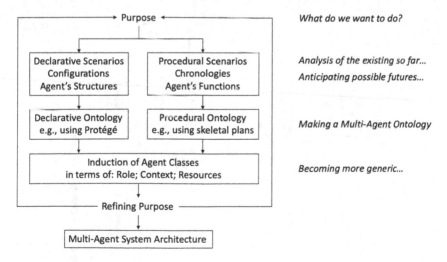

Figure 6.4 Making a multi-agent ontology that supports the development of a multi-agent architecture.

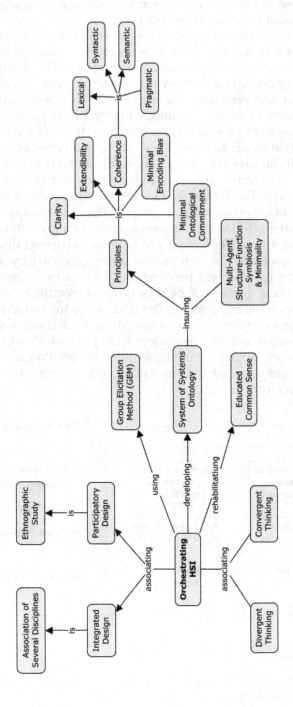

Figure 6.5 Orchestrating HSI.

The difference between the development of classical "single-agent" ontology and multi-agent ontology is the possibility of emergence of global (system of systems) behaviors and properties that force to refine roles, contexts, and resources, as well as purpose ultimately. Participatory design of such ontologies is mandatory.

Summary

Orchestrating HSI (Figure 6.5) consists in associating integrated design (i.e., associating of several disciplines mandatory for engineering design) and participatory design (i.e., ethnographic studies). It also consists in associating DiTh and CoTh. The GEM is usually a very good candidate to fulfill that association goal. Expertise and experience are very useful in orchestrating HSI, especially by rehabilitating educated common sense. Orchestrating HSI goes with the development of a system-of-systems ontology that insures principles (clarity, coherence, extendibility, minimal encoding bias, minimal ontological commitment) and multi-agent structure-function symbiosis and minimality.

Design for flexibility in an increasingly complex world

> Abandon the urge to simplify everything, to look for formulas and easy answers, and to begin to think multidimensionally, to glory in the mystery and paradoxes of life, not to be dismayed by the multitude of causes and consequences that are inherent in each experience – to appreciate the fact that life is complex.
>
> **M. Scott Peck**
> *Psychiatrist and writer*

Humans always tried to extend their natural capabilities by designing tools appropriate for a job. The more tools have become available, the more complex human business has become. At the same time, additional rigidity has been added in our everyday lives. Email is extremely useful but constrains us to read and write messages all the time. Can we say that this is natural? No, email introduced rigidity in our lives. Talking directly to someone is far more flexible. We can say that email provides flexibility in the sense that email is asynchronous, and therefore, we can write and read email whenever we want and can, as opposed to be constrained to move to a different location to talk to someone. However, we can also say that, because it is so easy to communicate, we are doing more than before, and complexity of communication tremendously increased because we now have huge quantity of emails to read and write. This example shows that the concept of rigidity, versus flexibility, is not straightforward and deserves more attention.

From the reductionist issue in engineering to risk analysis

The other day, I was discussing with one of my colleagues about the reductionist issue in engineering. He was telling me that reductionism was not an issue because we have built so many complex systems using linear algebra so successfully that he did not see any reason to process differently. Well! Reductionism in engineering and our current society in general leads to problems that are decomposed into smaller and smaller subproblems, which in turn leads to jobs that are increasingly specialized and most of the time disconnected from each other. Information technology enables "tighter and tighter control of ever smaller processes" (Schlundt, 2011). We have been immersed into this reductionist[1] syndrome forced by our engineering culture and training, which mainly involves deductive reasoning. We learned how to simplify a complex problem. More specifically, we learned how to reduce complexity to linear solvable problems. I remember that time when I learned linear algebra, when I had to manipulate matrices in order to find appropriate patterns that were easy to solve. This was a great exercise, like a game… I enjoyed doing it! Today, I think differently. I learned why we had to do engineering this way, the reductionist way. I learned that by reducing complex problems to matrices (like Excel spreadsheets), we could solve subproblems independently from each other, in the same spirit of the division of work – very little interdependencies. However, when we keep separating design processes of components that are unfortunately dependent, engineering solutions induce extremely complicated issues at operations time.

For example, when a new aircraft is being designed, it should be considered as a connected system within a system of systems, otherwise one day or another people who will be operating it will have to manage these interdependencies. This usually results, at operations time, in very rigid problem solving and sometimes failures. Sale issues of the Airbus 380 obviously belong to this category, at least partially. Airlines have difficulty in buying this two-deck aircraft just because too many airports are not equipped for loading and unloading passengers. Making a commercial aircraft is the job of an aircraft manufacturer, but interconnectivity among aircraft, airport infrastructure, and passengers should be considered at design time (i.e., considering the TOP model from the start). Who is doing this? Very few people in our market economy. It is like if there were as many composers to design a symphony as instruments in the orchestra (i.e., a composer for the violin, a composer for the percussion, and so on). How do you think the symphony would be at performance time? Instead, **a composer is an integrator who knows about all instruments of the orchestra and coordinates their plays to get a harmonious symphony.** In classical engineering, we expect too much – sometime subconsciously – from the human operators who are supposed to glue operations pieces in the end! Consequently, Human–Systems Integration (HSI) is, therefore, the way to go as early as possible during the life cycle of a system.

---

[1] Descartes's analytical thinking led to reductionism that consists in dividing a complex system into parts, analyze each part separately, and consider that the sum of the parts represents the whole initial system.

As already said, **becoming familiar with complexity** is often much better than simplifying complexity. How do we become familiar with a complex system? How do we become familiar with a new location or country? It takes time and efforts. Complexity is usually coming from a large set of components and an even larger sets of links among these components. Becoming familiar with such complexity requires that we identify what components really are and learn about links among them. A conductor needs to identify the various players and their instruments, as well as relationships among them in order to better understand their contribution to the symphony. He or she needs also to identify risks that a musician makes an error in order to keep the flow of the overall symphony.

Energy industry (e.g., nuclear, oil and gas) requires more development of **risk analysis** and anticipation methods and systems. It is always possible and more or less easy to explain why an accident happened, but it is very difficult and sometimes impossible to predict and anticipate it. Therefore, it is crucial to work more on prediction (i.e., causal deduction) and anticipation of possible futures (abduction). The best way to combine prediction and anticipation goals is to develop prevention experience feedback mechanisms that consider all defaults observed during operations, and integrate them into digital twins, which are subsequently used to support prevention, diagnosis, decision-making, and action taking. Failure and human error detection can be managed using sensors (i.e., what it is, where it occurred, what system [human and/or machine] was involved), information processing (i.e., using artificial intelligence [AI] to understand why it occurred, what are possible consequences), and projection engines (i.e., again using AI to infer possible plans and actions). For example, gas leaks could be detected, analyzed, and visualized to improve situation awareness and support decision-making and, finally, anticipate risk of dangerous gas cloud generation. Today, we are able to acquire large amounts of data, automatically analyze them (using appropriate AI algorithms), and infer relevant information that can be visualized in real time to human operators remotely located. Most advanced human–computer interaction (HCI) techniques and tools can be used to manage such data in order to support expert analyses. This is the reason why **operations rooms** should be thought very early during design and development processes to work on usages and not only on algorithms (without considering their usages).

It is clear that even if such an approach can be designed and developed in virtual environments, it also needs to be supported by realistic simulators (e.g., in the case of energy systems, a facility that includes tubes, valves, reservoirs, and other physical tools should be developed and used). Again, tangibility should be a main drive in such a risk analysis enterprise. Both virtual and physical simulation facilities should support normal, abnormal, and emergency situations, as close to the real world as possible. In addition, new systems could be tested using these facilities, such as sensors, information-processing systems, and even invent new jobs for the people involved.

Dealing with the unexpected as problem solving

No matter what you do in safety management, there will be a time when you will need to deal with the unexpected. Several authors developed concepts and approaches that describe and organize responses to unexpected events. For example, Taleb proposed the **Black Swan** concept.[2] He claimed that "Black Swan logic makes what you don't know far more relevant than what you do know. Consider that many Black Swans can be caused and exacerbated by their being unexpected" (Taleb, 2010). In this chapter, we will consider "dealing with the unexpected as problem solving." This view implies several requirements that include

- Training people to acquire problem-solving skills
- Developing technology that enables flexible crisis management
- Reshape organizations to insure safe, efficient, and comfortable management and outcome.

Let's talk about **risk taking**. In safety-critical environments, you may have taken all precautions, there will be a time when you will be facing with an unexpected situation and need to decide and act. In other words, you will have to take a risk. Risk taking is about preparation and routine task management (Boy & Brachet, 2010). Preparation is about mastering both domain and problem solving (e.g., it is highly recommended to climb many mountains in the French Alps before climbing the Everest mountain!) We also learned that routine is a killer. In other words, when you get too familiar with an environment, you tend to be complacent and decrease vigilance and underuse your attention skills. Mastering complexity in a given domain requires becoming familiar with this domain but also to keep being vigilant at all times.

Learning how to deal with the unexpected is a matter of preparation, attention, and appropriate reaction, with a possibility of failing… of course! We want to minimize this possibility of failing. We can improve human capabilities and also design appropriate flexible technology (this is what I call *FlexTech*) and incrementally develop human-centered organizations.

---

[2] "… in Europe, … swans are invariably white. Like purple cows and flying pigs, the black swan was a symbol of what was impossible. In medieval Europe, unicorns had more credibility. Dutch navigator Willem de Vlamingh, by finding black swans in Western Australia in 1697, showed how risky it is to declare something impossible" (Low, 2016).

This chapter presents an approach to deal with unexpected events by considering complexity, flexibility, maturity, autonomy (of machines and people), authority, as well as resource and context management.

Complexity and unexpected events

During the last three decades, we never stopped improving HCI within aircraft cockpits. We started with glass cockpit during the nineties, where we integrated many pieces of information in order to provide pilots with **meaningful information**, as much as possible in context. For example, in military jet cockpits, tactical situation is now presented in a more integrated manner (i.e., several data are fused to present meaningful information from a tactical point of view). In other words, we improved syntax to improve semantics management. It is like providing what a magician can do best (Tognazzini, 1993), removing syntactical tricks to leave the floor to the real object, that is tasks, targets, and more generally, mission.

In parallel, we also never stop developing digital systems that control mechanical things. We put layers and layers of software on top of each other to transfer cognitive functions from people to machines. In commercial aviation, we digitized aircraft to the point that we talk about "interactive cockpit" not because pilots interact with mechanical parts of the aircraft but because they interact on cockpit screens using a pointing device. **Digitalization** development contributed to remove several jobs and replace people by software-based systems. It also seems to be the only way to handle air traffic management (ATM) complexity (e.g., two programs were, and still are, extensively developed, SESAR in Europe and NextGen in the US, to deal with air traffic capacity increase). We concomitantly developed and installed more sensors. Some people talk about cyber-physical systems (CPS) and others about Internet of Things (IoT) to give a name to this growing set of systems that associate physics to information technology. At the same time, networking is the emerging concept.

All this is well, but we still have to consider people and organizations correctly in this evolution. We see more incidents and accidents coming from **unexpected events**. How can we anticipate unexpected events? How can we deal with unexpected events when they occur? Can we consider such unexpected events at design time and solve problems by design? This is what we will analyze in this chapter. But, even if we develop more barriers, we will always be facing unexpected events. This is the reason why we need to think about flexible technology that enables people and organizations to deal with such unexpected events correctly (i.e., safely, efficiently, and comfortably).

Dealing with the unexpected also means dealing with domain complexity. Complexity is about high interconnectivity. Interconnectivity is about nodes and links. When nodes and links become numerous, complexity increases and several phenomena arise such as feedback loops, emergence of new properties, attractors,[3] and potential catastrophes[4] (Thom, 1972). Complexity can be disaggregated in **meaningful contexts where procedural information can be concentrated into operating procedures that deal with expected situations**. However, information and knowledge remain to the discretion of people and organizations involved in systems management, which should have **specific skills and appropriate supporting technology to handle unexpected situations outside of these contexts**. These specific skills are about situation awareness, decision-making, and action taking (i.e., risk taking, most of the time). Appropriate supporting technology should enable flexible operations, not increase workload leading to stress, and support business continuity. Business continuity is about operations continuity, by opposition to operations rupture. It should be anticipated and programmed as both prevention of and recovery to potential breakdowns or disasters.

Designing innovative complex systems

When Louis Blériot crossed the English Channel for the first time on July 25, 1909, he took a calculated, well-prepared risk, as did, several years later, Jean Mermoz in the first flight between France and South America. Technological evolution has always involved a series of risks. From Eole, the flying machine designed and flown by Clément Ader in 1890,[5] to the Airbus A380, a multitude of decisions have been made so that each successive technology attains an adequate level of maturity. Taking a risk[6] means choosing among several options in order to move forward. Many accidents took place before aviation became the safest means of transport

---

[3] The term "attractor" is commonly used in mathematical theories of nonlinear systems that involves chaos. For example, the butterfly effect discovered and coined by Lorenz (1963). An attractor can be thought as a space where information or matter accumulates (e.g., a desktop is an attractor of all kinds of things related to work such as papers, pens, files, and books).

[4] Catastrophe theory is a branch of dynamical nonlinear systems mathematics and a specific case of singularity theory in geometry. It enables to explain unexpected behavioral changes where complex systems have been "abusively" linearized (i.e., using a too strong reductionist approach).

[5] Eole accomplished a very short flight of around 50 meters at a height of around 20 centimeters on October 8, 1980, at the Chateau d'Armainvilliers in Brie, France.

[6] In 2008, I organized a conference on risk taking within the framework of the Air and Space Academy. This conference gathered very highly skilled and knowledgeable risk takers in various industrial sectors such as aeronautics, space, nuclear, medicine, mountaineering, military, and intelligence services. We made a synthesis (Boy & Brachet, 2010) stating that "risk taking is a right, but also entails duties: that of anticipating its consequences, particularly on others, and that of proper preparation."

of the planet. We are now so used to flying without worrying about the risk that we often forget that there is no such thing as zero risk! In spite of this slight risk, the number of aircraft in European skies has increased by 4.5% per year on average in the past thirty years. Consequently, ATM complexity tremendously increased, and the crucial issue of congestion around large airports still remains unresolved. This issue forces us to look for new ways of improving safety and efficiency in the ATM by developing methods and tools to better analyze and understand increasing complexity.

Innovation and creativity are often a question of smart integration. How can this kind of integration be smart? It is smart when you know as early as possible how the resulting system will be used. Knowledge about usages is a key ingredient of innovation. How can you find out about possible usages when nothing or very little is built? There are two ways: (1) you anticipate possible usages (i.e., possible futures), develop related prototypes, test them in simulation, and assess how to modify your original ideas and concepts, and (2) you develop scenarios (i.e., simulation plans), run Human-In-The-Loop Simulation, and try to discover emerging usages, that in counterpart will help you modify your original plans. The former is decided, the latter is discovered. In both cases, validation is necessary using real-world scenarios.

Life-critical systems

Modeling and simulation capabilities are now ready to support both global and precise design infrastructures and functions. They enable the development of flexible applications that can provide quick results in uncertain environments. This is what is needed in human-centered design (HCD) of life-critical systems (LCSs) today.

What do we mean by life criticality in engineering design? This is what it takes to design systems when human life is on the line. An LCS is anything that contributes to support our everyday life whether at work or our personal life. LCSs deal with safety, efficiency, and comfort. We sometimes realize that a system is life-critical when it disappears. For example, imagine that you lose your smartphone! What do you do? You cannot call the phone provider to inform that you lost it. You suddenly realize that there is a lot of data that you don't want other people to access, and so on. Smartphones have become LCSs. Indeed, smartphones brought us autonomy, but in counterpart, they brought hyper-dependency also. They are like prostheses that have become necessary in our everyday life. The more we use them, the more they become prostheses because we lose our previous capabilities of managing our lives without them.

At this point, the complacency issue needs to be introduced. The more we use LCSs, the more we become complacent to what they are providing to us. Problems come when we continue to trust them outside of their context of validity. This is a way to define risk. Risk is the likelihood of a specific effect within a specified context, which is in fact the outside of the context of validity of the system at stake. Risk could be expressed as a complex function of probability of occurrence of an event combined with its consequences and vulnerability. For example, the probabilistic risk assessment (PRA) method quantifies risk as an expected loss:

$$\text{Risk} = \text{Sum}(CS_i * P_i)$$

where P_i is the probability (likelihood) of occurrence of each consequence (i.e., the number of occurrences or the probability of occurrence per unit time) and CS_i is the consequence severity, which is the magnitude of possible adverse consequence(s) (e.g., the number of people potentially hurt or killed).[7] HRA forces you to answer the following questions:

- What can go wrong with the system? (i.e., what are the initiators or initiating events (undesirable starting events) that lead to adverse consequence(s)?)
- What and how severe are the potential detriments? (i.e., adverse consequences that the system may be eventually subjected to as a result of the occurrence of the initiator?)
- How likely to occur are these undesirable consequences? (i.e., what are their probabilities or frequencies?).

[7] In a previous book, I developed a case (e.g., Fukushima Daiichi nuclear power plant disaster in Japan) where CS_i was very large (i.e., close to infinity) and P_i was very small (i.e., close to zero) (Boy, 2016). Indeed, mathematics say that zero multiplied by infinity is undetermined. The problem is intrinsic to the probabilistic risk model. It would be much better to choose a "possibilistic" approach to risk. Using possibility theory (Dubois & Prade, 2011), we learn that probability is a single number that does not consider ignorance (or degrees of ignorance). Instead of representing uncertainty of an event "E" by a probability "$P_i(E)$" (i.e., a number between 0 and 1), it is much better to choose an interval, with an upper value, called "possibility" (i.e., a number between 0 and 1, and noted "$Pos_i(E)$"), and a lower value, called "necessity" (i.e., a number between 0 and 1, and noted "$Nec_i(E)$"). Formally, we have the following relationships: $Nec_i(E) \leq P_i(E) \leq Pos_i(E)$. The ignorance "$Ign_i(E)$" that we have on event "E" is given by the difference: $Ign_i(E) = Pos_i(E) - Nec_i(E)$. Total ignorance is: $Ign_i(E) = 1$. This means that: $Pos_i(E) = 1$ and $Nec_i(E) = 0$. Therefore, instead of having a single formula [$R_i(E) = P_i(E) \times CS_i(E)$], we will have two. First, we will assess the possible risk using the following formula: $R_{Pos,i}(E) = Pos_i(E) \times CS_i(E)$. Second, the necessary risk will be: $R_{Nec,i}(E) = Nec_i(E) \times CS_i(E)$. Of course, in the case of the Fukushima Daiichi disaster, since $Pos_i(E) = 1$, what only counts is $CS_i(E)$ and necessity, which could be calculated using factors and data provided by civil engineering and nuclear engineering in particular. We will not enter into any calculation, but we can say that $Nec_i(E)$ is bigger than zero. I also strongly advise the reader to get more familiar with the introduction to the new statistics proposed by Cumming and Calin-Jageman (2017), to understand effect sizes, confidence intervals (CIs), and meta-analysis ("the new statistics"). In any case, critical thinking is key.

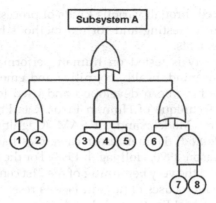

Figure 7.1 Fault tree analysis based on a "and/or" gates decomposition.

At this point, an event tree analysis and a fault tree analysis should be carried out. A fault tree analysis is based on a hierarchical decomposition of entities linked among each other by "and/or" logical gates (Figure 7.1). It is performed in five steps:

- Define the undesired event to study
- Obtain an understanding of the system
- Construct the fault tree
- Evaluate the fault tree
- Control the hazards identified.

The main question is: what do you do when a system fails? There are five types of LCSs:

- Fail-operational systems continue to operate when their control systems fail (e.g., elevators, gas thermostats, home furnaces, passively safe nuclear reactors) – they may be unsafe
- Fail-safe systems become safe when they cannot operate (e.g., medical infusion pump, domestic burner controller, nuclear weapons, railway signaling)
- Fail-secure systems maintain maximum security when they cannot operate (e.g., electronic doors that lock when there is power failure – in contrast with fail-safe doors that unlock) – they may be dangerous
- Fail-passive systems continue to operate in the event of a system failure (e.g., in case of an aircraft autopilot failure, aircraft remains in a controllable state) – manual reversion is the key issue
- Fault-tolerant systems avoid service failure when a fault is introduced into the system (e.g., nuclear reactor control systems, BITE [Built-In Test Equipment], normal maintenance failing systems replacements).

LCSs are engineered through various kinds of processes such as CMMi[8] and product maturity testing and formal methods based on legal and regulatory requirements.

Human reliability is tested on human performance in relation to age, errors, state of mind, health, cognitive and emotional stability, for example. Methods have been developed and used for several decades, such as THERP (Technique of Human Error Rate Prediction), COCOM (Contextual Control Model), and CREAM (Cognitive Reliability and Error Analysis Method). Human error categories have been developed (Norman, 1981; Reason, 1990; Hollnagel, 1998). For the last two decades of the 20th century and the very beginning of the 21st century, human errors were one of the main focuses of human factors research and practice – a negative role of humans! For the last decade or so, resilience engineering has become prominent, promoting a positive role of humans. "Success has been ascribed to the ability of groups, individuals, and organizations to anticipate the changing shape of risk before damage occurs; failure is simply the temporary or permanent absence of that."[9]

Risk taking is often omitted from textbooks presenting risk – such a taboo! However, any time you have taken all precautions to execute a dangerous task, you will have to act at some point – you will have to take the risk to act. Doing nothing is an act that could be as dangerous as doing something! We developed this question with a panel of risk takers going from experimental test pilots to astronauts, mountaineers, surgeons, special force agents, lawyers, nuclear power plant operators, and company CEOs (Boy & Brachet, 2010); all risk takers of this panel agreed on the fact that preparation was the key factor for risk taking – routine is a killer (i.e., when everything is going perfectly well, people become complacent). As a matter of fact, we take risk for various kinds of purposes that include discovery, challenge, helping others, and just for pleasure (Figure 7.2).

We may take risk by ignorance and management of uncertainty. This is the reason why probability theory is not sufficient to manage risk. We need to handle risk using other kinds of uncertainty and imprecision theories (interval theory, fuzzy sets, etc.).

Risk taking can be voluntary and involuntary. This changes everything. This is the reason why experience, knowledge, and skills are so important. You cannot take a risk without knowing what you are doing. Again, this is related to preparation. Risk taking deals with crisis management (i.e., problem solving and experience).

Risk taking is also related to laws and regulations, which are related to a compilation of experiences and debates, often informal, between

---

[8] Capacity Maturity Model integrated.
[9] www.resilience-engineering.org/.

Figure 7.2 Are there actions that are not risky?

people who have seen risk close enough to talk about it intelligently and "carefully." Of course, there is always a first time, and expertise is again crucial by analogy with other situations where people have risk taking experience. Think about Neil Armstrong who landed on the Moon for the first time "by hand" without any help from automation.

Back to separability, from an HSI perspective, people's capabilities are important to consider. Indeed, as David Llewellyn, from the School of Sport and Exercise Sciences of University of Leeds, UK, finds out when he was studying mountain climbers' attitudes related to risk taking, there are three types of people (Figure 7.3):

- Risk avoiders
- Risk reducers
- Risk optimizers.

At this point, we saw that instead of reducing complexity to simpler paradigms, we prefer to become familiar with complexity and analyze risks related to the use of the complex system being considered. However, we can make our complexity analysis job easier by figuring out what parts of the system of systems can be studied in isolation from the rest. Llewellyn's classes of climbers constitute a good example of criteria separating the complex system of mountain climbing into three sub-systems where problems to be solved become more affordable. More specifically, the tryptic technology, organization, and people as a system of systems is easier to investigate. Requirements can be developed along with these three human orientations, for example.

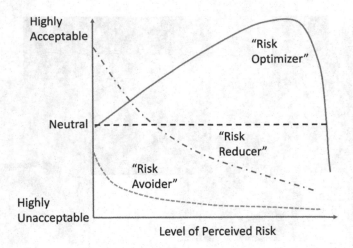

Figure 7.3 Possible relationships between level of perceived risk, risk orientation, and risk acceptability (Llewellyn, 2003, p. 27).

Designing for flexibility requires appropriate socio-cognitive representations

People involved in a difficult and complex problem-solving activity should be as open as possible (i.e., they should not be constrained to follow rigid rules and procedures at all times). Indeed, the procedure following is very effective and safe when the context covered by procedures is very close to the real-world context where they are used. However, outside of this context, they should be equipped with specific skills and appropriate flexible technology.

What is flexible technology? In a world where almost all machines will be automated, flexible technology, or FlexTech, will support flexible operations. People are the only systems (if systems include people and machines as already defined previously) that can effectively make strategic decisions and solve problems in a flexible way. This is the reason why we need to further develop intelligent assistant systems (Boy, 1991a). Obviously, this statement is not new since we already developed semiautomated systems that people can handle in a flexible way. It is not because you can develop a very advanced automaton that you will provide people with a useful, usable, and performant system in the end. People and machines should co-adapt.

If, as already discussed, flexibility is a matter of co-adaptation, then modeling co-adaption is crucial. For a long time, cognitive engineering was based on the information-processing system model (Figure 7.4), where an information processor and its knowledge base grinds information coming

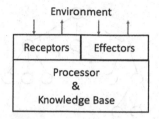

Figure 7.4 System as information processor.

from sensors and produces information sent to effectors. Receptors and effectors interact with the environment of the system.

However, the information-processing system model is too limited since it is single agent (i.e., a human or a machine facing its environment). A multi-agent approach is, therefore, more appropriate because it is about interaction among agents (Figure 7.5). In addition, an agent is a society of agents (Minsky, 1986). This recursive definition of an agent is fundamental because each level of the representation is multi-agent. In that sense, as already described, a multi-agent system is a system of systems.

Each agent is an organization (i.e., internal organization of the agent system of systems) and belong to an organization (i.e., external organization to the agent as a system within a bigger system of systems). The organization of a system of systems may be based on various kinds of models. The hierarchical organization often comes to mind because it reflects the way conventional organizations work in our social world (Figure 7.6).

Information flow in a hierarchical organization is mostly top-down with very little information going upstream. Therefore, coordination is structurally insured (i.e., hierarchical organization structure determines

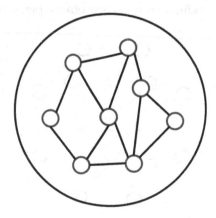

Figure 7.5 An agent as a society of agents.

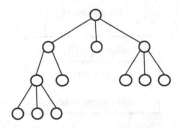

Figure 7.6 Hierarchical organization of a society of agents.

the way coordination will be handled). Conversely, a heterarchical organization will require more ad hoc coordination rules considering not only structural setups but also functional contexts. For example, road rules have been defined with respect to various infrastructure patterns, such as roundabouts, road crossings, and highway entrance. There are cultural differences in road rules with respect to countries. In France, for example, when two roads merge, if you come from the right, you have priority. This determines car driver behavior. In addition, police makes sure that road rules are well followed to insure safety.

The orchestra organization is an alternative to hierarchy. In an orchestra (Figure 7.7), each agent is an expert performing an equally complex task. However, in contrast with the hierarchical organization where agents either control other agents and/or report to another agent, each orchestra agent acts with respect to scores that have been coordinated by a composer, and a set of rules shared by all agents of the society of agents (i.e., coordination rules). These rules are defined using specific syntax and semantics. For instance, music theory provides a set of such rules for music players in an orchestra. In addition, the overall coordination is managed by a conductor, and there are transversal communication channels (e.g., each musician is connected to the conductor and listens others' performance).

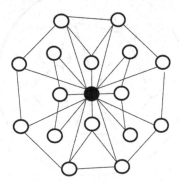

Figure 7.7 Orchestra organization of a society of agents.

In the ATM domain, for example, the more aircraft become autonomous, the more they should become knowledgeable and informed agents about their surrounding neighbors and natural obstacles. By autonomous, we mean that they know where they are located and what their environment is about (situation awareness) and where they need to go with respect to the current and future situation (goal orientation). Up to now, airspace hierarchical organization was necessary and sufficient because aircraft were not numerous and not autonomous from a navigational point of view. Today (and tomorrow), this configuration is changing. For example, most aircraft are now equipped with global positioning systems (GPSs) that provide more autonomy (Stix, 1994). The GPS provides pilots with a more precise location of the aircraft; thus, pilot's situation awareness is theoretically improved. Therefore, they are in a better position to decide by themselves (i.e., they are more autonomous). This statement is valid if they trust the GPS. The hierarchical concept is no longer compatible with this evolution. In contrast, the orchestra concept is more appropriate. Each aircraft is an agent. Control centers are other agents that work as conductors. The difficulty with several possible conductors is guidance coordination and consistency.

Following up on the orchestra metaphor, the conductor makes sure that each player plays the right music at the right time with the right mood to contribute to the harmony of the symphony. Note that these rules can be explicit (e.g., written on scores) or implicit (e.g., induced by conductor's gestures). In the shift from hierarchical air traffic control (ATC), where air traffic controllers dictate what pilots must be doing, to increasingly heterarchical ATM, where "air traffic managers" make sure that coordination rules are followed, in the same way policemen make sure road rules are followed on roads. Technology brings more autonomy to agents, which requires providing them with more sophisticated situation awareness, decision-making, and action taking capabilities. Such capabilities associated with such coordination rules cannot be acquired in one shot but should be incrementally constructed.

Designing for flexibility requires appropriate socio-cognitive representations. System-of-systems (i.e., multi-agent) representation is the solution that is proposed in this book, considering that a system is the basic representation of technology, organizations, and people (refer to Figure 4.1).

System-of-systems flexibility

The question now is characterization of systems, as agents, and links among them. Each system, whether a human or a machine, has at least a function characterized by a role, a context of validity, and a set of resources to achieve function's role (refer to Figure 4.10). Note that function's role is (or should match) the (generic) task that the function must perform. Links between two systems can be of various kinds: supervision, delegation, cooperation, trust, and so on. System-of-systems flexibility comes from the fact that some internal systems can be modified, removed, or replaced, as well as some links can be modified, removed, or replaced both internally and externally.

Each agent has a specific task to perform. Usually a high-level task is assigned to the system of systems at the top level. Since a system of systems is recursively defined (Figure 4.10), it has several levels going from cognitive at the top to physical at the bottom. We say that a system is more cognitive than another when it involves more cognitive tasks and activities, that is, more information processing (thinking).

For example, let's assume that you are at work and your car does not work (this is context), your high-level task is "going back home." At this point, it is a goal leading to a high-level cognitive task that needs to be implemented (i.e., the goal should be decomposed into sub-goals, and so on). In practice, you go to the bus stop, wait for the bus to arrive, get into the bus, pay the driver, wait in the bus until your descent at the right stop, walk to your home. This is called a plan! Very much rigid, isn't it? What about flexibility? Let's assume that a friend of yours drives close to you walking to the bus stop. He opens his car window and addresses you by saying: "How are you doing? Are you going home? Would you like a lift?" At this point, you may say: "Yes, why not!" At this very moment, you break your initial plan following up an unexpected event. From a cognitive viewpoint, you switch **from a goal-driven cognitive process** (i.e., intentional behavior) **to an event-driven cognitive process** (i.e., reactive behavior). You then enter into your friend's car, and you restate your initial plan. You are telling your friend where to go. You become a navigator.

What does systemic flexibility involve? Remember that a system is defined by its structure of structures and function of functions. Considering the previous example, you, as a flexible system, should be able to easily allocate appropriate functions when a specific event occurs. This also assumes you have these appropriate functions for such changes and are able to control and coordinate them. In addition, the system should be capable, and sometimes authorized, to allocate functions in various kinds of situations, such as normal, abnormal, and emergency situations. Coordination rules should be stated and used. The intentional/reactive flexibility is a human asset, but it should be better thought in the machine area.

> Flexibility is the exercise of free will or freedom of choice on the continuum to synthesize the dynamic interplay of thesis and antithesis in an interactive and innovative manner, capturing the ambiguity in systems and expanding the continuum with minimum time and efforts.
>
> *(Sushil, 1997, 1999, 2000)*

Flexibility is, therefore, associated with authority, responsibility, and domain capability. **Authority** can be given to a human or machine system (i.e., a priori assigned authority) or emerge from recognition (i.e., experience feedback shows that a system should be in charge). For example, astronauts learn both **leadership** and **followship**, because they could be a priori assigned to lead a task but have to support somebody else performing another task. In the future, with the integration of AI, there may be situations where a human is in charge, but when the situation changes, to become abnormal or emergency, the machine could take the lead and vice versa. People and machines should be prepared to deal with such change situations.

Social contexts and maturity of technology and practices

Social representations (SRs) influence the way technology is used. Even if we know that SRs guide action, a lot needs to be explored to better understand how in a given context (Lahlou, 2004). Saadi Lahlou developed a theory of **cognitive attractors** (CAs), complementary to SR theory. CA is a micro-psychology theory focusing on the moment when the subject, as a result of interpreting his present context, takes actions. A CA is a set of physical and cognitive items that potentially participate in a given activity and are simultaneously perceived by the subject. People interpret the attractor as an activity Gestalt and is automatically driven into the corresponding activity. The attractor's strength is the combination of several factors: pregnance (attraction of attention), cognitive cost, and motivational value.

Unexpected situations may emerge from the use of systems that are not yet mature. It is then important to better understand what we mean by maturity. As already explained, we make a distinction among various kinds of maturity: technological maturity, maturity of practice, and societal maturity.

In this chapter, it is time to better explain what maturity is about. Following up on Henri Bergson's quote (at the beginning of the chapter), to exist implies to change that implies to mature... for example, cars never stopped to change and, therefore, mature from the beginning of the 20th century, improving safety, efficiency, and comfort. Their technological maturity contributed to modify types of drivers, going from experts to anybody for regular cars, to specialized drivers for specific activities (e.g., bus drivers, Formule-1 drivers). An entity, whether human, organizational, or technological, that cannot regenerate itself endlessly will never mature and die.

Maturity follows a cycle from birth to death of any entity. There are three periods:

1. The entity creates all possible attributes to insure maturity during Period 1 (i.e., entity is dependent of its organizational environment – how much it helped learning, for example)
2. The entity uses these attributes when mature during Period 2 (i.e., entity can autonomously handle its own existence in its environment)
3. The entity dies when its attributes become obsolete during Period 3 (i.e., entity cannot autonomously handle its own existence in its environment any longer).

There is still an immense field of research to explore on the influence of CAs on the three types of maturity. Think about gesture interaction used on smartphones and tablets that attracts people to behave in a (now) standard way almost anywhere anytime.

System-of-systems autonomy

Let's consider the airspace as a system of systems (i.e., a multi-agent system), where systems are increasingly autonomous and interact among each other (Figure 7.8). In the following, we analyze how the various agents interact along with the three models of interaction already described (Figures 4.11–4.13). These models are context-sensitive (i.e., an agent [i.e., a system] should be able to change from supervision to mediation to cooperation according to context). Consequently, they enable flexibility.

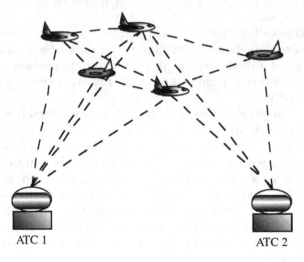

ATC 1   ATC 2

Figure 7.8 ATC centers and aircraft as a system of systems (i.e., multi-agent system).

In practice, a technological, organizational, and/or human **autonomous system of systems** has skills and knowledge that enable it and all its constituting systems to survive in the largest number of situations possible.

Supervision is usually supported by procedures and checklists. In aviation, operations procedures are usually learned and remembered by pilots. Checklists are implemented and used to extend the human short-term memory. They are guides used to maintain people on the "right track."

Mediation advocates the use of a (party line information[10]) common database, managed by specific agents or systems commonly called mediators, to improve situation awareness.

Cooperation advocates appropriate mechanisms that enable to facilitate construction of a mental model of agent's environment. This mental model also includes mental models of the other agents involved in the overall interactions, within the system of systems.

All three models of interaction have in common the fact that situation awareness should be shared among interacting agents. In the supervision model, the supervisor should be aware of what executants are experiencing. This could be done via reporting or direct observation. In the mediation model, the common database should reflect shared situation awareness. In the cooperation model, the more agents are familiar with one another, the more they share a common situation awareness.

Summary

Design for flexibility in an increasingly complex world (Figure 7.9) requires risk analysis, innovation, and dealing with the unexpected (i.e., training people in problem solving; having flexible technology for crisis management; reshaping the organization to obtain safe, efficient, and comfortable management and outcome – basically insure business continuity). Risk analysis uses methods, contributes to develop safety-effective systems, and performs risk-taking analyses based on various risk takers' profiles. Design for flexibility also requires socio-cognitive representations that are multi-agent capable of survival at all, or as many as possible, levels of granularity and easily switching from goal-driven to event-driven strategies and should consider social context, technology maturity, and maturity of practice.

---

[10] Aircraft pilots use shared voice VHF frequencies to increase their situation awareness with respect to other aircraft and environmental conditions. This is called "party line" information (PLI) in the aviation community (Hansman, Pritchett & Midkiff, 1995). PLI has been a controversial topic when we introduced datalink systems on board aircraft because of PLI loss risk.

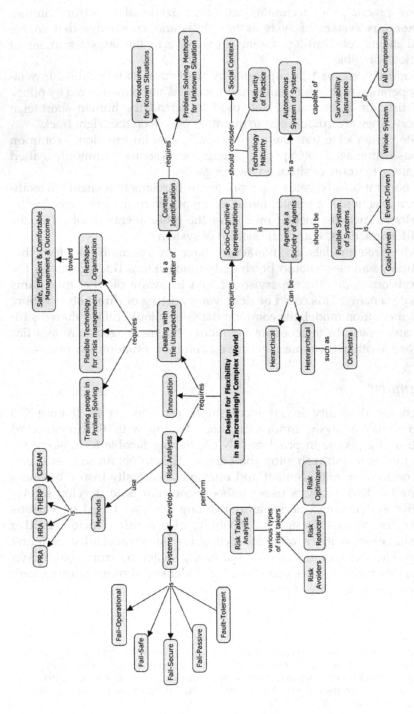

Figure 7.9 Design for flexibility in an increasingly complex world.

chapter eight

Activity, creativity, storytelling, and tangibility
The right mix

> The whole difference between construction and creation is exactly this: that a thing constructed can only be loved after it is constructed; but a thing created is loved before it exists.
>
> **Charles Dickens**
> *Writer and Social Critic*

Virtualization (we also talk about dematerialization) happens everywhere today, opening boundaries of creativity (Cadoz et al., 2014). Cadoz and his colleagues are working on digital technologies for computer arts and computer music. They are considering embodiment, presence, enaction, and tangibility seriously and relate these concepts to creativity, within the European Art-Science-Technology Network (EASTN). They described "tangibility" in terms of reality, materiality, objectivity, presence, concreteness, and so on. In this chapter, I expand these views to Human–Systems Integration (HSI).

If creativity is a matter of openness, in life-critical systems such as aerospace, it does not come without expertise. This is the reason why I decided to associate creativity to storytelling and tangibility for the construction of innovative systems. Storytelling is provided by experts (e.g., airline pilots). Tangibility comes from incremental Human-In-The-Loop Simulation (HITLS) of implemented stories of the future (i.e., incremental validation of anticipation).

You may learn swimming theoretically for a while, but you will need to jump in the pool to learn swimming! Anytime you have to take a risk, you need to prepare yourself as much as you can, but there will come the time when you will have to act. This is the same in design. You may perform lots of formal developments, and I claim that both natural/artificial versus cognitive/physical (NAIR) and structure/function versus abstract/concrete (SFAC) models, presented in the introduction of the book, will help figuring out how to state your design problem, and later on solve it, but you will need to figure out how to start making concrete system design at some point and evaluate it.

This chapter is devoted to the articulation of several concepts and analysis methods related to task, activity, performance, workload, timeline, and other human-centered design (HCD) approaches. We will also present innovation approaches.

Timeline analysis: from taskload to workload analysis

In the beginning of the 1980s, I was involved in the certification of the first two-crewmen cockpits for commercial aircraft. We were moving from three to two aircrew members and had to show that safety and efficiency would be as good as it usually was. We started by eliciting knowledge from pilots that we recorded on timeline sheets (Figure 8.1), which included sequences of elapsed times (determined by the flight plan, which was decomposed into phases and sub-phases, and so on), event descriptions, task descriptions, human agents involved and their actions (CM1 for crew member 1, typically the captain; CM2 for crew member 2, typically the first officer; and CM3 for crew member 3, typically the flight engineer who was eliminated on new-generation commercial aircraft), and possible comments.

Once timeline sheets were informed, we had a time framework determined by elapsed times that provided available times t_A. Each task had a required time t_R. This was useful for the calculation of taskload (TL) based on the following mathematical model:

$$TL = t_R/t_A$$

A digital model of the aircrew, cockpit, and ground control, called MESSAGE,[1] was developed and used to compare the two-crewmen cockpit to the three-crewmen cockpit both quantitatively using TL

Elapsed Time	Event Desc.	Task Desc.	CM1 Action	CM2 Action	CM3 Action	Comments

Figure 8.1 Timeline analysis sheet.

---

[1] MESSAGE was for "Modèle d'Equipage et des Sous-Systèmes Avion pour la gestion des Equipements" (which meant "Model of crew and aircraft sub-systems for equipment management").

measures and qualitatively using comments provided by subject-matter experts (Boy, 1983; Boy & Tessier, 1985).

This very simple method provided very interesting results that were compared to data coming from other workload metrics based on physiological measurements and subjective evaluation scales (e.g., Cooper–Harper handling qualities rating scale [Figure 2.2] – Cooper & Harper, 1969). Typically, pilots learn how to assess their own workload, based on the Cooper–Harper rating scale; at the same time, potential observers rate their perceived pilots' workload using the same scale. These ratings were performed in flight and/or in simulator. A possible implantation is the following: each pilot is facing a set of ten aligned buttons and an alarm light that is activated by an experimenter; the pilot is required to turn the light off by pressing a button representing his/her current workload.

Later on, other methods were developed such as the NASA TLX (Hart & Staveland, 1988). These methods belong to subjective evaluation techniques. They require specific formalisms and foremost models that are often controversial. Workload is a difficult concept that is both a product (i.e., people produce workload by performing work) and a physical and cognitive input for human operators (i.e., nothing can be done correctly without a positive arousal). It is, therefore, a regulation variable internal to all human beings. The main question is, how much workload do we need to perform well? In fact, there are two thresholds for workload: an upper threshold of stress and a lower threshold of vigilance. Best workload profile is always located between these two thresholds, not too high and not too low!

Studying workload of air traffic controllers (ATCOs), Jean-Claude Spérandio showed that changes in ATCO's strategies appeared attributable to workload variations (Figure 8.2). ATCOs modulate their workload and strategies to keep optimal performance (Spérandio, 1984). Subsequent

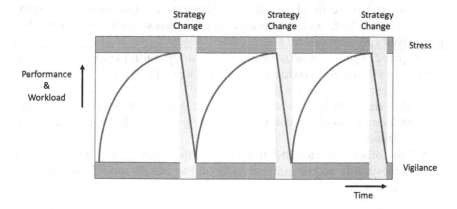

Figure 8.2 Variations of performance and workload between vigilance and stress. (Adapted from Spérandio, 1984).

research in air traffic control showed that "... workload auto-regulation rests on the interval of time between a diagnosis (or real suspicion) of conflict and the moment when a definite solving action is undertaken. Modulation of the length of this time interval is the basic mechanism through which controllers can regulate their workload" (Athènes et al., 2002). This time interval has been called the maturing time (MT) that tends to increase attentional level and, therefore, workload but decrease decision-making uncertainty. Consequently, compromises have to be found between uncertainty and time pressure.

Activity analysis in the form of storytelling

For a long time, activity analysis is an asset of the French-speaking association for ergonomics (SELF[2]). We already saw the difference between task (i.e., what is prescribed to be done) and activity (i.e., what is effectively done). There is still a controversy between two meanings of activity:

- (M1) activity as a product produced by a human transforming a task into an activity
- (M2) activity as a function that transforms a task into something different that can also be called activity.

In both cases, we can observe resulting behavior (i.e., what we "see") and actual performance (what is effectively produced and not always necessarily observable). Behavior and performance may not be the same, since behavior is observed directly and performance often needs to be "reconstructed" from observable data. Observing behavior is subjective, as performance metrics are often constructed from objective measures. In this book, I adopted meaning M1 following the definition of a human function (cognitive or physical) that transforms a task into an activity. As a matter of fact, such a human function corresponds to meaning M2 (i.e., activity as a function).

Figure 8.3 presents activity results as an operating compromise from the conflict between two different logics (Hubault, 1995):

- One directed by the task, the technology, and organization logic that corresponds to the artificial side of the NAIR model (i.e., what is asked).
- The other directed by the human (operator), the life logic that corresponds to the natural side of the NAIR model (i.e., what it requires).

---

[2] Société d'Ergonomie de Langue Française.

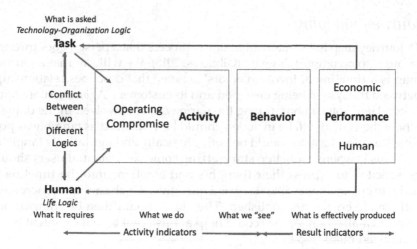

Figure 8.3 Relating task, activity, behavior, and performance. (Adapted from Hubault, 1995).

- A distinction is made between activity indicators and result indicators. This distinction is very useful when we observe people at work to elicit activity, behavior, and performance for further cognitive (and physical) function analysis (Boy, 1998).
- In design, it is crucial to show potential users of the system being designed its performance and advantages. This is extremely difficult in the beginning of the design process when the system does not exist yet. This is the reason why shared visualization of ideas, drawings, mockups, and prototypes is a need. In addition, you should involve these potential users in the design process in order to get their subject-matter expert participation. These two requirements, **visualization** and **participatory design**, should then be integrating parts of your HCD method. Finally, it is very difficult to figure out if your customers will use the system being designed sustainably without knowing them well and what they want to do. Consequently, you need to work with them on scenarios that prefigure possible futures and test these possible futures with them. In other words, you should help them constructing where they want to go (i.e., you should help them give birth to their future system – this is called Socratic maieutic[3]). Journey mapping is a method that enables carrying out such Socratic maieutic.

---

[3] Socratic maieutic is a mode of inquiry, which aims at bringing a person's latent ideas into clear consciousness. Somebody practices Socratic maieutic when he or she helps another person to give birth to targeted appropriate ideas.

Journey mapping

"A journey map is a visualization of the process that a person goes through in order to accomplish a goal" (Gibbons, 2018). We will say that a journey map is a **timeline** of involved actors' actions[4] that describes relationships between the system being designed and its customers. Again, visualization is key. Here, interactions among the various actors involved in the domain where the system will be included should be visualized as tangibly as possible (i.e., visualization should be both physically and figuratively tangible).

This is where and when **storytelling** comes in. Potential users should be solicited to express their thoughts and emotions into the timeline in order to create a believable story or a narrative, which should be incrementally made crisper and polished. The story should then be transformed into a visualization. Artists could help at this point to make content being visualized more credible.

This is also where the **TOP model** comes in. Interactions among technology, organization, and people should be visualized in a tangible way (i.e., physically and figuratively credible).

Journey mapping helps a design team acquire what kind of **user experience (UX)** the system being designed induces. It enables to figure out this UX by observing potential users' goals and motivations. You can observe learning difficulty and help these potential users to improve their understanding of how to use the system. You can observe human errors and help potential users improve their abilities to recover from these errors and/or modify the system in order to avoid such human errors to happen again.

Such journey mapping practice helps HCD of the system being designed. Resulting participatory design very much helps consolidating relationships between design team, manufacturing company, and potential users. For example, the following questions to potential users should help the design team to better understand potential users' requirements and, consequently, formulate them correctly. Why potential users are interested in the system being developed? What would user's learnability curve look like? How easy is the system to be put at work without too many changes in the organization? Is the system efficient and effective? How extendable this journey map is if we need to scale up? Answers to these questions will help you better understand potential users' motivations and goals.

Like for in classical timeline methods, a user journey map represents the journey of a person (i.e., an actor or stakeholder in the system being designed), including the scenario he or she follows (or will follow) expressed in terms of phases, his or her expectations and goals, opportunities, and internal ownership. Table 8.1 presents major components of a journey map.

---

[4] The term "actor" is commonly used in journey mapping. It means "human agent," to be consistent with the term "agent" used in this book.

Table 8.1 Major components of a journey map

Actor	Generic user (persona) experiencing the journey (e.g., pilot or first officer in the aviation example above).
Scenario and expectations	Scenarios can be existing or anticipated (e.g., two- or three-crewmen cockpit). Expectations are to insure safety and efficiency.
Phases	Phases are contextual segments in the journey (e.g., a flight is usually composed of phases such as takeoff, climb, cruise, approach, and landing).
Actions, mindsets, and emotions	Actions are user's behaviors. Mindsets include user's thoughts, questions, motivation, and information needs during the journey. Emotions are "ups" and "downs" of UX (e.g., user can be delighted or frustrated).
Opportunities	Opportunities are insights gained from journey mapping development (e.g., What needs to be done with this knowledge? How are we going to measure improvements we implement?).

Journey mapping will not be further developed within the scope of the book. Many consulting companies are using this method, and a simple search on the web will show you commonalities and differences of the various implementations. I strongly suggest that you try; again, "jump in the pool to learn swimming!" In any case, "journey mapping is a process that provides a holistic view of user experience by uncovering moments of both frustration and delight throughout a series of interactions. Done successfully, it reveals opportunities to address customers' pain points, alleviate fragmentation, and, ultimately, create a better experience for your users" (Gibbons, 2018). One more thing: It takes empathy to run journey mapping sessions!

A few words about innovation[5]

What would be human life without innovation? People and organizations need to innovate to regenerate themselves. However, we need to be aware of the following issues:

- Humans and organizations are naturally conservative (i.e., they may protect existing practice and disregard new ideas)

---

[5] This section is mainly based on discussions with my colleague Jérémy Legardeur, who is the creator of the "24 Hours of Innovation," and Julien Ambrosino's PhD work on innovation, as well as my own experience in the aerospace domain.

- Processes of successful innovation are collective, even if design and invention can be individual
- Structure of the innovation team or organization should be adaptable and scalable and at the same time integrate all necessary contributors
- Strategy for institutional leadership that not only influences innovation but also is influenced by innovation.

We already discussed the duality between complexity and familiarity. Innovation could be described as **transforming a complex problem into a familiar solution** (i.e., a solution that will easily become familiar at operations time).

Innovation is often rejected because people fear uncertainty in effectively producing new concepts, breaking usual standards, and dealing with social issues such as making some jobs obsolete. Innovation is an economically viable invention (Schumpeter, 1911, 1942).

Innovation requires a flexible approach understandable by all stakeholders. It should induce motivation and as less constraints as possible. Again, flexibility and meaning are key factors.

Product (or system) performance evolution from inception to maturity can be represented by an "S-curve" such as in Figure 8.4 (Foster, 1986; Christensen, 1992; Mann, 1999). Each new product evolution can be represented by an S-curve that can be compared to another one toward the selection of the best one or merging of several.

We have used similar curves in aeronautics where criteria were not only performance but also safety, efficiency, and comfort (Figure 8.5). These curves are inverted compared to S-curves because we considered that functions represented damages. Note that damages asymptotically

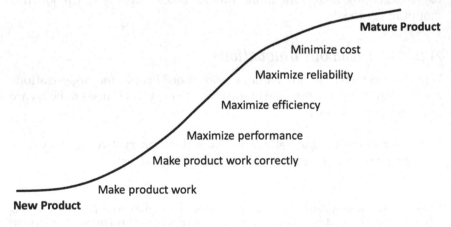

Figure 8.4 S-curve (Mann, 1999).

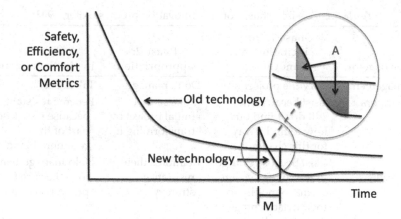

Figure 8.5 Evolution of damages due to safety, efficiency, and comfort over time (Boy, 2011).

tend to zero. Curve representing new technology should be under the previous one. Segment M represents the maturity period.

Region A represents the acceptability area, as the sum of relevant parameter values, for example, number of casualties, bad performance, or discomfort situations, over the maturity period M. The identification of appropriate M and A thresholds is crucial to assess if a technology will succeed or become obsolete according to contemporary acceptable practices and lifestyles. In addition, it is important to identify the best time t_D for final product delivery and actual operations start. If this time t_D is within the maturity period M, we usually talk about surprises when unacceptable events occur. When t_D is after the maturity segment M, there are still residual unacceptable events, but they are more due to routine practice and complacency than to unknown events or behaviors.

Other methods, more quality assurance oriented, were developed such as Quality Function Deployment (QFD), which was initiated and developed in Japan in the late 1960s and introduced to America and Europe in the early 1980s (Akao, 1997). QFD method consists in transforming operations demands into design quality (i.e., deploy functions forming quality) and deploying methods to insure design quality and further manufacturing processes in what we call today a system of systems.

Innovation can lead to the renewal not only of the object of innovation but also of the playground (i.e., ecosystem where it is introduced) and induces several kinds of uncertainty and unknown (Alter, 1993). This is represented in Table 8.2, where the organizational actors such as management, innovators, and legal specialists do not have the same kinds of issues.

Table 8.2 Possible phases of the innovation process (Alter, 1993)

Type of actor	Phase 1: from incitation to innovation	Phase 2: appropriation	Phase 3: institutionalization
Management	Carry the project	Don't manage	Takes over
Innovators	Resist because they still don't find out how to find a way for this change	Make sense of initial project by transforming it	Become resistant because they lose a part of the conquered land
Legal specialists	Resist because change in the ruling system seems to modify to their disadvantage the established order	Reinforce their resistance strategy	Help management to make sure to put in place rules

The evolution of innovation is organized into three phases: from incitation to innovation, appropriation, and institutionalization.

Ambrosino (2018) proposed a list of major innovation skills: creativity, autonomy, experimental spirit, ability to identify opportunities, integrative skills, ability to process data from a variety of disciplines, ability to adopt different points of view, critical thinking, ability to formalize problems (precise specifications), taste of concretization, customer logic (integrate it into your design team), attractiveness for the future and change (do not try to fight resistances but adopt a positive approach), ability to establish links, and ability to create a sense of shared development.

Innovation will not be further developed in the scope of this book. Just remember that innovation is a difficult exercise that is not only limited to ideation but involves rigor, engagement, and regularity as well. An invention without proactive market analysis and incremental prototyping and testing (i.e., agile development) will barely qualify as innovation.

Do not forget to make things tangible…

Mixing creativity, storytelling, and tangibility correctly means that you have good ideas, you can tell them correctly, and you can show a prototype. Tangibility is both physical and figurative (we can also say "cognitive"; both terms are interchangeable within the scope of this book). This means that you need to visualize (i.e., the physical part of tangibility) and explain (i.e., the figurative part of tangibility) what the system is about; how it works; and, from a HSI point of view, show

how technology, organizations, and people are involved in the system. Showing tangibility does not wait; do it as early as possible using modeling and simulation.

From an HSI perspective, tangibility may not be visualized and explained in the same way all the time. For example, an aircraft manufacturer that is designing and developing a new airplane will show various kinds of personnel's advantages to an airline. It is a B-to-B[6] approach of HSI. Conversely, a computer company that is designing and developing a new laptop will show to customers (i.e., everybody) various kinds of advantages for them directly. It is a B-to-C[7] approach of HSI. However, if the aircraft manufacturer is developing a passenger cabin, the HSI approach will be B-to-B-to-C. Consequently, HSI testing requires the right stakeholders (e.g., end users whether involved in operations or maintenance).

Summary

Finding the right mix of activity analysis, creativity, storytelling, and tangibility is still more an art than a technique (Figure 8.6). However, several techniques are provided in this chapter. It requires timeline analysis, activity analysis (e.g., storytelling, observation in HITLS or at work, journey mapping), innovation, measuring maturity, and business tangibility.

---

[6] Business to business (a term commonly used in marketing).
[7] Business to customer.

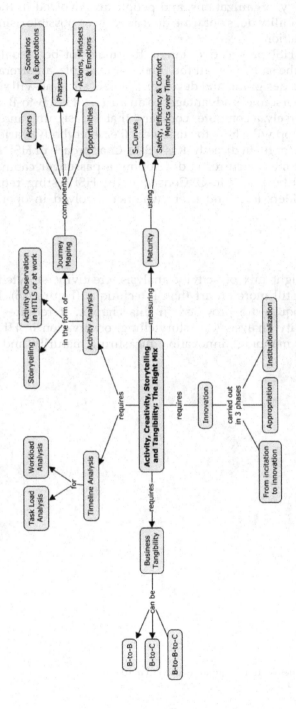

Figure 8.6 Activity, creativity, storytelling, and tangibility: the right mix.

chapter nine

Evaluation processes and metrics

> We've all heard it, 'You can't manage what you can't measure.' But the problem with metrics-driven leadership is that it doesn't work. When you try to manage by the numbers, ... you drive towards mediocrity. Quantitative measurements alone will never make an organization great, because it is the qualitative elements of performance that achieve greatness.
>
> **Lisa Earle McLeod**
> *Sales Leadership Expert*

Measuring human and organizational factors is not an easy task. It is sometimes even impossible to define the right metrics. At the same time, when we accumulate expertise and experience in a domain, we know when something works and when it does not work. Difficulty is not so much to solve a problem but to state it right. We often miss the point by willing to measure human and organizational factors quantitatively, while we do not have the right metrics, which can only be qualitative most of the time. In this sense, Lisa Earle McLeod is right when she promotes qualitative elements. In this chapter, we will review evaluation methods and metrics useful for Human–Systems Integration (HSI).

As already discussed in the introduction of the book, HSI can be seen as an extension of existing disciplines that are human factors and ergonomics (HFE) and human–computer interaction (HCI). Both fields already developed methods and metrics that can be used. HSI extends and uses them in systems engineering (SE) during the whole life cycle of a system. HSI is at the intersection of several disciplines that include HFE, HCI, and SE, to which we can add artificial intelligence (AI), architecture, and other engineering disciplines more appropriate to the application domain being covered.

The main contribution so far is the *Handbook of Human Systems Integration*, published by the American Psychology Association (Boehm-Davis, Durso & Lee, 2015). In this chapter, a few more conceptual and practical tools[1] are provided. Now, before starting to assess or measure

---

[1] Some of these conceptual and practical tools are being developed within the FlexTech Chair that the author coordinates.

anything, it is crucial to understand what are the goals that we want to achieve by building a new system and constraints that we have to achieve these goals, as well as in what context evaluation will be performed.

What do we mean by metrics?

Metrics should qualify, quantify, measure, and benchmark some kind of performance. HSI metrics are focused on the evaluation of human–machine systems.

Any metrics is based on a model or theory. Therefore, models that support metrics described in this chapter are also presented. Evaluations, in the form of verification and validation, are typically conducted during the design and development process (i.e., we talk about formative evaluation) and before delivery (i.e., we talk about summative evaluation or certification).

Evaluations require appropriate metrics related to technology, organizations, and people (the TOP model again!). We need to keep in mind that HSI requires a systemic approach that does not only look at TOP components individually and independently but also and foremost focuses on TOP integration (i.e., investigation of system-of-systems performance).

Complexity of a system of systems induces emergent properties that are not included in its parts or components. In addition, complex sociotechnical systems always evolve (i.e., sub-systems can be added, removed, or modified dynamically). Therefore, HSI will have to be evaluated at different levels of **granularity** and in various **contexts** of operations. The concept of context has been defined in Chapter 4. Consequently, creativity, prototyping, and incremental evaluations are complementary processes that could be implemented in an agile manner.

Let's talk about goals and high-level metrics

Why should we build a new system? For example, let's assume that we want to build a new aircraft that satisfies a major issue, that is, air traffic congestion. Indeed, air traffic increases by an average factor of about 4.5% every year for the last forty years. The growth is exponential. If we do not find appropriate solutions, air traffic complexity will cause chaotic effects very soon, in the sense of complexity science (Mitchell, 2009). Three possible solutions emerge from the analysis of this problem: build bigger airports (hubs), build bigger aircraft, and automate the sky (e.g., this is the purpose of programs such as NextGen in the USA and SESAR in Europe). Since the second option, that is, build larger aircraft, seems reasonable, it could be considered as a goal for an aircraft manufacturer, let's say building an aircraft that doubles or triples the capacity in terms of passengers.

What kinds of metrics would you use to evaluate the feasibility of such an aircraft? My suggestion is to use the TOP model and take metrics for aircraft technology, metrics for related organizational aspects (during design and development, as well as during operations), and metrics for people who will be involved during the life cycle of the aircraft. This necessarily obliges to consider the aircraft itself and its environment (e.g., airports) as a system of systems.

More generally, you will need to consider high-level metrics that will be used as a filter to evaluate the system being designed and developed (e.g., aircraft) as a connected object. These high-level metrics are threefold: social, environmental, and economic. Sustainability is at the intersection of these three factors in a given sociotechnical culture (Figure 9.1). This is also precisely how the United Nations define sustainability.[2]

The economic sector together with political sector are about market dynamics that determine constraints and objectives evolution and development and control strategy.

The social sector is about work and operations conditions and regulations, as well as communication and cooperation modes. HFE and HCI associated with SE are handling this sector, leading to human-centered design (HCD).

The environmental sector already set up sustainability metrics. An example of work on environmental metrics for sustainability is included in a Columbia University report (Cohen et al., 2014) and summarized in Table 9.1.

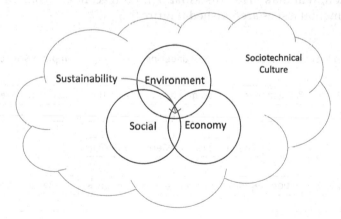

Figure 9.1 Sustainability at the heart of economic, environmental, and social high-level factors in a given sociotechnical culture.

---

[2] www.unenvironment.org/about-un-environment/why-does-un-environment-matter/environmental-social-and-economic.

Table 9.1 Example of sustainability metric

Social	Private sector
	• Human rights and resources
	• Performance in products, production, and supply chain
	Public sector
	• Safety and health
	• Population
	• Infrastructure
	• Budget and expenditure
	• Education
Environmental	• Energy
	• Emissions
	• Disclosure
	• Water
	• Materials
	• Effluents and waste
	• Biodiversity
Governance	• Transparency
	• Equality and fairness
	• Efficiency
	• Corruption

Tangibilization of Virtual HCD, which is the result of systemic HCD (i.e., an association of HFE, SE, and HCI) carried out using modeling and simulation, will **make HSI sustainable** if both economic constraints and environmental issues are satisfied (Figure 9.2).

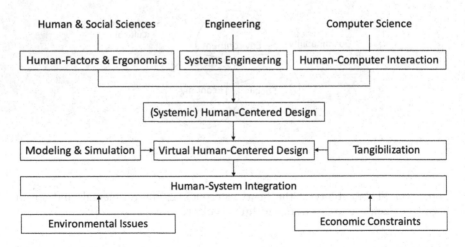

Figure 9.2 Making HSI sustainable by using HFE, SE, and HCI and satisfying economic constraints and environmental issues.

Another example is a classification of economic, environmental, and social factors that was recently studied and described in vehicle loading and routing operations (Vega-Mejia, Montoya-Torres & Islam, 2016).

Now, if we apply this approach to the example of the large aircraft whose goal is to contribute to solve the air traffic congestion issue, here is what I could suggest. HSI high-level metrics could be as follows (this is just a preliminary list, very far from being completed):

- Social: safety, security, health, efficiency, comfort, control, workload, etc.
- Environmental: fuel consumption, noise, carbon/CO_2 trace, etc.
- Economic: costs, sale prospective, equality, intramodality benchmarking, connectivity (airport compatibility), etc.

The "human" component of HSI is clearly considered as interconnected with his/her social, physical, and economic environment.

Analysis, design, and evaluation

HSI should be started as early as possible during design and development project. Virtual HCD was already promoted to provide a playground for activity (i.e., user experience) observation and analysis, formative evaluation, prototype refinement, and iteration until a satisfactory solution be found (Figure 9.3).

We already saw sketching in Chapter 8. It includes ideation, storyboarding, development of mockups going from cartoon-like drawings to very sophisticated animations, and simulations. Engineering design follows by developing computer-aided design (CAD) models and specifying dimensions in more detail. At this stage, we definitely move from

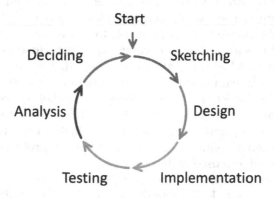

Figure 9.3 HSI iterative design and development cycle.

art to engineering. Then comes implementation in the form of a kind of videogame (e.g., using Unity-3D) that can be tested in the next phase of the cycle. Testing involves Human-In-The-Loop Simulations (HITLSs), various kinds of scenarios, relevant criteria and metrics, and appropriate test users. Recorded test data are analyzed with respect to established principles (i.e., HSI and domain-specific principles). The last phase of the cycle is decision-making to determine whether a new iteration should be started or not.

It is highly recommended to use the **AUTOS pyramid** to avoid forgetting essential principles (see Chapter 3). Indeed, AUTOS has been developed and tested to this end during the last three decades. I advise to be systematic and use all edges of the AUTOS pyramid to develop metrics, measurement models, and observable variables (see Chapter 2).

Depending on the maturity of system development, evaluation can be analytical (in the beginning), formative (during the development process), and summative (at the end to certify the system).

At each stage, **physical and cognitive function analyses** should be performed to update HSI knowledge and check that maturity is coming along smoothly (i.e., technology maturity, maturity of practice, and social maturity). Obviously, function analysis starts by a task analysis; functions are "discovered" along the way and incrementally added in the functional ontology of the system being designed and developed. As already presented in Chapter 8, timeline analyses and journey maps greatly help defining tasks and functions. At the same time, emerging functions can be allocated (or reallocated) to systems and sub-systems structures. This is also a way to effectively determine realistic levels of automation. The more the function of functions network become mature (i.e., stable from one iteration to the other), the more human jobs and machine algorithms can be finalized. When it is the case, summative evaluations can be started and finalized toward certification.

HSI iterative process should be cost effective as well as effective in terms of safety, efficiency, and comfort. How many design and development cycles should we plan? Is it possible to answer this question from the start? Well, this is a question of **diverse expertise and experience** in both HSI and domain when the system is developed. This means that design and development team members (i.e., designers, engineers, human factors experts, HCI specialists, and other appropriate stakeholders) should cooperate and share the same objectives. In addition, HSI information is not only a matter of a priori user requirements elicited from HFE specialists but also knowledge coming from HITLS on virtual prototypes and digital twins developed and used at design time.

Evaluation is typically done using methods such as failure modes and effects analysis (FMEA), fault tree analysis, usability engineering

(Nielsen, 1993), cognitive task analysis, field observation and ethnography, workload assessment, time and accuracy of users' performance, risk analysis, user satisfaction ratings, human error analysis, subjective scales, fatigue assessment, situation awareness (SA) assessment, and event data analysis. This is a non-exhaustive list adapted from a list proposed by Pew and Mavor (2007).

Evaluation can be done by direct observation (on the field), analysis of recorded work scenes and other experimental data, interviews of human operators and experts, questionnaires, brainstorming (or brainwriting) with an appropriate group of end users and experts (i.e., participatory analysis, design, and testing), and so on. In aviation, flight tests in the real world or simulation are always a great source of useful data for activity analysis.

Types of metrics

Evaluation depends on the type of phenomena that we are trying to investigate. In Chapter 2, a first account of **tangibility metrics**, measurement models, and observable variables has been provided. They include: (1) intrinsic complexity of the system being developed and extrinsic complexity of the environment influenced by the system being developed; (2) technological maturity, maturity of practice, and societal maturity of the system being developed; (3) passive and active stability, robustness, and resilience of the system being developed; (4) flexibility expressed as ease of operations provided by the system being developed in abnormal and emergency situations; and (5) sustainability influenced by and influencing economy, environment, and society. Links with lower-level HSI methods were also provided in Table 2.1.

In this chapter, let's provide more methods and tools that cover the entire AUTOS pyramid framework (i.e., the ten edges of the pyramid):

- Task and activity analysis (U–T)
- Information requirements, and technological requirements and limitations (T–A)
- Ergonomics and training (procedures) (T–U)
- Social issues (U–O)
- Role and job analyses (T–O)
- Emergence and evolution (A–O)
- Usability/usefulness (A–S)
- SA (U–S)
- Situated actions (T–S)
- Cooperation/coordination (O–S).

Task and activity analysis

Task analysis was used for a long time for writing training manuals. More recently, it is used up to front in HCD and business process analysis. We should make sure **at design time** that the system being designed, whether technological or organizational, is useful, because later is too late! First things, first! This requires us to understand what activity will be when people will use the system (i.e., what will be the job of users involved). We already discussed the importance of the distinction between task (i.e., what should be done) and activity (i.e., what is effectively done). It is then important to combine task analysis and activity analysis through proto-type testing (Figure 9.4).

There is a variety of task analysis methods developed and used in HFE and HCI (Palagi et al., 2018; Kanki, 2018; Patrick & James, 2004; Jonassen, Hannum & Tessmer, 1999; Hackos & Redish, 1998; Desberg et al., 1986; Watson, 2003; Calvary et al., 2003; Kirwan & Ainsworth, 1983; Paternò et al., 1997; Drury, 1983).

In our digital world, let's take HCI[3] task analysis approach in terms of introspection, questionnaires, ex situ and in situ questionnaires, and laboratory experiments.

Task and activity analyses should be done at the right level of granularity. In addition, the level of granularity may vary with respect to context. Task/activity knowledge can be expressed in the form of scenarios, Unified Modeling Language (UML) use cases and task models.

The iBlock formalism (i.e., interaction blocks) was proposed as a task model (Boy, 1998):

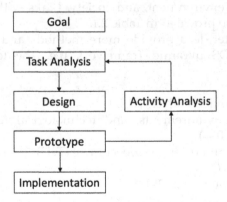

Figure 9.4 HCD process involving both task analysis and activity analysis.

---

[3] Human–computer interaction (HCI) is the discipline that studies design, implementation, and evaluation of useful and usable human-centered interactive systems.

- Triggering conditions
- Context
- Algorithm of tasks (usually a procedure)
- Normal final conditions or goal
- Abnormal final conditions.

This definition shows the recursive property of the concept of task (i.e., a task is a task of tasks). At the lowest level, a task is a cognitive or physical action. For example, a cognitive action can be a memory access to remember something or processing of an observed information and a physical action a mouse click or a maneuver on a joystick.

For example (Figure 9.5), you want to produce a letter – this is the goal. The original context is what is available to you, such as a computer, a text-processing application, and a printer – these are resources. Producing a letter requires to edit and print the letter. Editing the letter requires type and modify text. Typing text requires moving the cursor to a specific place on the screen, select the right keys, and so on. Tasks are decomposed into simpler tasks that lead to executable actions.

iBlocks can be linked among each other forming a network. For example, an iBlock goal or abnormal final condition usually leads to a triggering condition of another iBlock. Several iBlock could clustered into the same context. iBlock clustering is very important because it enables testing complexity of the various activities involved in the system of systems being designed and developed.

iBlock task model departs from GOMS (goal, operators, methods, and selection rules) (Card et al., 1983) by introducing the concepts of abnormal conditions, which is very useful in the design of life-critical systems. It also expands by enabling to cluster iBlocks into meaningful contexts. In addition, GOMS enables handling sequences of actions only (i.e., no parallelism and more generally algorithms). Other formalisms are available such as HTA (hierarchical task analysis) or UAN (User Action Notation).

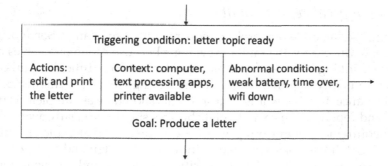

Figure 9.5 Example of an iBlock for "Producing a business letter."

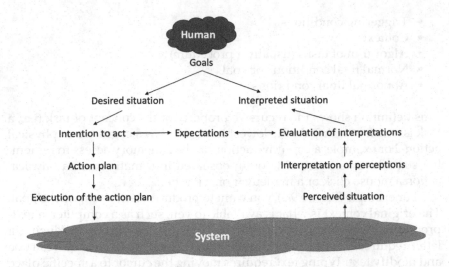

Figure 9.6 The "action cycle." (Adapted from Donald Norman, 1986).

Implementation of the HCD process (Figure 9.4) requires support of an appropriate activity model. Activity analysis is performed based on the active cycle described on Figure 9.6. More generally, the discrepancy between the prescribed task and the activity (i.e., effective task) is studied step by step to determine if the planned task is realistic or should be modified to fit real-world constraints and requirements.

Whenever a goal is expressed, it can be translated into a desired situation that can lead to an intention to act and then an action plan that can be executed on the system. The system can be observed and a perceived situation is formed, interpreted, and assessed. An interpreted situation is formed when a reasonable assessment of the expectation of the difference between the desired situation and the interpreted situation is minimal.

Technological requirements

During the life cycle of a project, requirements can change. Some priorities may differ from the initial previsions because economy evolves and design thinking is refined in terms of implementation difficulties, missing or contradictory requirements, and other things related to unanticipated performance. In other words, requirements should be as complete as possible and repercussions of potential changes should be clearly available.

Requirement engineering is a crucial process that typically establishes TOP dimensions: technology in terms of design and costs, organization in terms of schedules and verification plans, and people in terms of skills and operational procedures. From a cost perspective, we already saw in Chapter 1 that technology-centered engineering V-Model looks like a check mark (Figure 1.2), where costs are lower in the beginning of the V and much higher in the end because not enough attention is brought to HCD that should include modeling and HITLS from the beginning of the design process. The cost of fixing errors is often much higher than putting enough resources that must be put in the early stages of the life cycle of a project to help **anticipate tangible operationality** (i.e., you need to find out operational problems as early as possible). Therefore, tangible operationality should drive requirement engineering, and configuration management should absolutely include HSI specialists. This is the reason why HSI departments should be in charge of such tangible operationality by gathering system engineers and the various stakeholders (what I call people in the TOP model) and documenting both ongoing solutions and their design processes (see section on "virtual HCD as document management" in Chapter 1). Design cards as presented above should be used, as well as the traceability mechanism attached to them.

Requirements need to be decomposed in finer grain requirements until manageable pieces are found. This top-down decomposition is crucial. At each step, it should include stakeholders' expectations and feasibility of proposed solutions. Prototyping helps figuring out how to turn requirements into systems. At each step, structures and functions should be tangible physically and figuratively (i.e., understandable). By simulating all possible parts and their interconnectivity among them, you will anticipate not only how the overall system will work and be used but also how to integrate it in a tangible way. Among many things, this means that you should put in place a mechanism that enables you to **verify** that parts that may have been bought from third parties will satisfy the requirements. Maturity metrics can help you verify both technological integration and also HSI (i.e., maturity of practice).

SE already provides techniques and organizational setups for technical management. HCD and HSI should be included in all processes including requirement engineering, technical planning, interconnectivity, and interoperability management (interfaces between processes and technical parts); configuration management; risk management; and data management. In addition, incremental assessment of roadmaps (e.g., design cards) and technical decision-making should be concretely organized and run frequently (i.e., in an agile way).

One thing that is specifically important is interface management that deals with technical interconnectivity and interoperability. Indeed, decomposing a system into sub-systems and so on is often meaningful in engineering terms but not at all in operations terms. We realize, often too late, that people are needed to articulate some processes and technical parts. HSI dictates to take care of this as early as possible by allocating appropriate functions to humans and machines.

Ergonomics and training

In this book, let's consider that ergonomics[4] is a mix of human factors and design (Tillman et al., 2016). We consider that ergonomics contributes by design to decreasing human fatigue and stress, and increasing human comfort and productivity. We usually make a distinction between physical ergonomics and cognitive ergonomics.

There are several domains under physical ergonomics[5]:

- Anthropometry that provides body sizes in a population; several software programs are available, such as ANTHROPOS and RAMSIS (human-solutions.com), DELMIA Safework (delmia.com), and ErgoForms (ergoforms.com)
- Guidelines for these postures (e.g., standing, sitting, reaching, moving)
- Work and living spaces (e.g., organization of offices, work stations, cockpits, stairs, accessibility, habitability, climate)
- Devices used for control (e.g., equipment, input devices, displays, consoles, and panels)
- Coding and alarms (e.g., use of size, shape, color, texture, location, labeling, sounds, verbal messages).

Cognitive ergonomics[6] focuses on human perception, mental processing, and memory. It is fairly developed since the 1980s. Today, cognitive ergonomics provides recommendations to designers on how people interact with digital technologies, apps, wearables, and IoT (Internet of Things), for example. These kinds of interaction are getting more intensive and demanding, both cognitively and emotionally. Cognitive ergonomics promotes intuitive interaction with digital media.

Cognitive ergonomics is interested in improving knowledge on cognitive systems (i.e., systems that have perception, information processing, and memory in them), life-critical systems (i.e., systems that involve safety, efficiency, as well as cognitive and emotional comfort), and complex dynamic systems (i.e., systems that are difficult to understand and may change their structures and functions more often than expected). Cognitive engineering is the engineering version of cognitive ergonomics (Norman, 1982, 1986; Boy, 2003).

Cognitive engineering looks for human–machine co-adaptation. This co-adaptation may take a substantial amount of time, and best indicators to measure it are related to time, assuming good performance.

---

[4] The word ergonomics comes from two Greek words: ergo (i.e., work) and nomos (i.e., laws).
[5] See Guidance Notes on the Application of Ergonomics to Marine Systems (2013–2018).
[6] The European Association of Cognitive Ergonomics (EACE) organizes conferences since 1982.

On the training side (i.e., people's adaptation to machines), indicators will be focused on human operator's physical and cognitive skills and knowledge.

Social and cultural factors

Analysis of social and cultural factors requires an architecture model, which entails socio-cognitive and ethnographical evaluation methods and metrics. At the center of this architecture is the concept of social agent[7] or system, as defined in this book. This architecture is focused on decision-making and its impact on all components (humans or machines) of the system of systems being studied. Such an architecture was proposed for the analysis of impact of various kinds of decision-making actions in the context of crisis management (Boy, 2010). An ontology of these decision-related factors was derived.

Most important social and cultural factors and associated processes are useful to first qualify and further quantify in job evolution, change management, collaborative work, accident prevention, crisis management, and participatory design, for example.

When we shifted from three-crewmen cockpits to two-crewmen cockpits, circa 1980–1982, removing the flight engineer from the flight deck, many pilots were not agreeing 100% with this evolution, which can be called a revolution. Automation partially replaced the flight engineer. In fact, we shifted from control to supervisory control (Sheridan, 1984). The captain and the first officer had to supervise engineering systems that were fully handled by the flight engineer before. If the evolution of technology seemed linear, it took us some time to realize the real human-centered revolutionary shift: we moved from control to management. This was not only a question of technology evolution, it was a drastic change in flying an aircraft.

For example, digital technology invaded not only flight decks but also progressively the entire aircraft. The aircraft itself became an agent in its own right. I like the metaphor of an aircraft as a horse. When people shifted from horses to automotive machines (i.e., cars), coach jobs became chauffeur jobs. Coaches had to understand horses' behaviors to get safe, efficient, and comfortable rides. This kind of task required expertise and experience. Chauffeurs had to learn how to steer (i.e., how to relate steering wheel rotations to car wheel rotations). This was straightforward in a sense (e.g., you turn the steering wheel to the right and the car will turn to the right), which was not necessarily the case for coaches steering the right

---

[7] Note that we often talk about "social actors." Typically, social actors are people immersed in a social context, with various kinds of factors such as individual, cultural, influence and socio-cognitive factors that determine their behavior.

rein expecting that the horse would turn right – the coach had to understand "horse's psychology," that is, horse's mood and reflexes. In other
words, coaches had to master horse's mental model. Now, shifting from
a mechanical vehicle to a highly automated vehicle is the opposite exercise. Drivers or pilots have suddenly to understand the mental model of
the automated vehicle. The more a vehicle becomes automated and gains
autonomy, the more people who have to manage this vehicle will have to
master their mental model (i.e., their possibly complex reactions to specific
inputs). Now, imagine the complexity of understanding system-of-systems
activity without the support of an appropriate architectural model.

What are social and cultural factors that emerge from our current
automation/autonomy evolution? Such factors emerge from various kinds
of adaptation of human and machine agents interacting between each
other to maintain an acceptable level of stability and safety. People usually
adapt much better than machines on complex tasks. Machine can adapt
on simple tasks using algorithms based on control theories and machine
learning, but this is limited to specific contexts and capabilities. In any
case, multi-agent adaptation requires coordination mechanisms that have
to be put in place. Back to the orchestra metaphor, both composers and
conductors are required to provide coordination support: composers at
the score level (i.e., the task level) and conductors at the performance level
(i.e., the activity level). Again, this requires expertise and experience for
both types of coordination.

Following up Minsky's definition of an agent (i.e., an agent is a society of agents), a social agent can be a group of acquainted human and
machine agents (e.g., a team of agents who know each other very well), an
organization (i.e., a larger group of people interconnected through rules,
often hierarchical relationships), or a community (i.e., a group of people
who share same interests). As already described, an agent can be modeled
by an architecture (i.e., a structure of structures) and a function of functions (see Chapter 4). A function is represented by three attributes: role,
context of validity, and a set of resources that support it operationally.
Function analyses enable to incrementally improve function allocation
among human and machine agents. Functions can be distributed among
various groups of agents.

Functionally driven interaction among agents facilitates the emergence of social objects. Claude Rivière introduced the concept of **social
object** in his essay of sociological epistemology (Rivière, 1969). Three
main issues are at stake to better define social objects:

- Current research techniques must be improved by adding two kinds
  of adaptation, that is, context-sensitive and motivated adaptation
- Validity of methods being used must be improved to insure
  objectivity

- Research rationale must be clearly defined (i.e., must elaborate social object epistemology).

Social objects are defined by their objectivity (i.e., they must be described impartiality and universality as much as possible). They are usually the result of a consensus among a group of specialists instead of coming from a single expert, for example. The best we can do is usually relative universality because we cannot get all opinions from all centuries, countries, experts, and so on. Objectivity can be partially obtained by

- Precision (e.g., through measures, axiomatization, predictions)
- Explanation (e.g., from understood facts, by seeking causes, thanks to a methodological pluralism, by inducing laws)
- The "total" approach (e.g., through historical anthropology) (Macfarlane, 1977).

Whenever we describe a social object, we have a part of subjectivity coming from our own culture and society considered as a norm. It is very difficult to make sure that we reach universal objectivity. This is why several experts and consensus reaching methods are required. However, we are always in a catch-22 situation where the object depends on the method and the method depends on the object. Any social object will be described as we perceive it. Obviously, this description should be useful for further prediction and evolution of interconnected social objects.

A social object emerges from purposeful interactions among agents. There is no social object without action. The social object does not exist as an isolated entity, but it is understood as a network of relations. Considering the cognitive function analysis approach (Boy, 1998), a social object is a persistent network of cognitive functions distributed among a set of agents (e.g., the network of mail cognitive functions distributed among a set of postmen). This is the reason why it is crucial to identify the **social context** in which this cognitive function network emerges. This network is characterized by a set of **individual, influential, and sociocognitive factors** that need to be elicited in order to explain resulting **collective behaviors** (Figure 9.7).

People tend to "economize" their lives by finding out patterns that enable them to simplify their interrelations among each other. In complex systems and environments, simplicity is always a matter of familiarity (i.e., frequency and regularity). Regularity tends to establish stable relations in groups. Paradoxically, the identity of a social object is only found when there is a sufficient diversity of subjects who share a common understanding of it, for example, the identity of a country. Yugoslavia's identity, for example, was only maintained for many years by a common deliberate social object called communism, which broke into parts

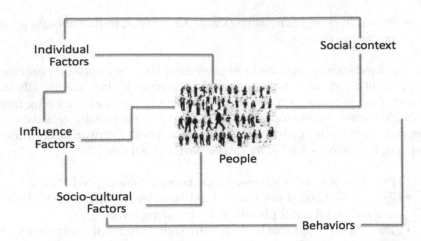

Figure 9.7 Factors that help describing social objects (Boy, 2010).

when the supervision model became obsolete and constituent identities came back (e.g., Serbia, Croatia, Bosnia, and Kosovo). A social object may emerge from several attributes such as religions, traditions, languages, and political beliefs.

More generally, social context is typically defined by cultural factors, beliefs, social evolutions, ability to cope with social changes, social norms, traditions, rituals, myth, customs, social values, moral, and obviously history.

Role and job analyses

Jobs are defined by a set of articulated tasks. Task analysis has been already presented. Job analysis should determine

- What people involved should be doing (i.e., their role in terms of task to be performed)
- What kinds of tools or systems they should be using (i.e., resources that they should use)
- What they must know (i.e., the type of knowledge that they must have to determine in which context they are; knowledge could be declarative, such as objects and relationships, or procedural, such as task sequences, goals, and sub-goals; in addition, knowledge includes dependencies and constraints).

Let's take factors that help describing social objects displayed in Figure 9.7. Agents in a social context have their own individual differences with respect to the following factors: personality, mental features, physical constitution and characteristics, affectivity, individual behavior, cognitive capacities, and social involvement. Some of these factors could be common to several social objects, while others are able to discriminate social objects among each other. It is important to elicit them to figure out useful distinctions that enable the identification of emergent social objects.

Influential factors can be developed in the same way. People can be influenced either by themselves or by others. Therefore, influential factors may be internal such as expectations (either people expect too much and their expectations are not fulfilled, or have wrong expectations, for example), accomplishment needs and motivation. Influential factors may be external such as extreme determination due to social pressure, manipulation, and history of other people (stronger people may influence weaker people, for example), cultural or religious influence, and public opinion via media and propaganda communication.

Sociocultural factors include cognitive dissonance (e.g., when people do not reason or process information in the same way), egocentrism (i.e., some people think about themselves and do not care about the others), social identity (e.g., depending on their social status, people are likely to produce different behaviors), orthodoxy (i.e., people obey strictly to a doctrine), social perception (i.e., perception of the social context may not be objective but rather subjective, the elicitation of such subjectivity is always useful), bias (i.e., we see with our biased eyes depending on our country, political beliefs, and so on), social regulation (i.e., other people reflect on us, experience feedback is constantly reinjected), social representation (i.e., the way we represent our society or another society depends on the representation that we use), and stereotype (e.g., French people shake hands or kiss women any time they meet).

Resulting behavior is precisely a social object. Accumulation of factors described above may end up in various kinds of behavior such as discrimination (i.e., special treatment taken toward or against a person of a certain group; it could be based on class or category; discriminatory behaviors usually induce exclusion, racism, or rejection), specific social relations (e.g., crime watching, over insurance), and specific status (e.g., specialization of groups of people, delegation of responsibility, authority sharing, accountability).

Let's further define this anthropological approach that supports the incremental definition of social objects. The proposed method is based on the search for types of interaction among agents using knowledge and know-how elicited from domain experts.

Systemic interaction models (refer to Chapter 4), which agents typically use, depend on the quality and quantity of knowledge that they have of the organizational environment where they interact among each other. There are various levels of interaction in a multi-agent system. At the local level, each agent should be able to interact with his/her/its environment in the most natural way. Actions means should be as affordable as possible at the right time and, more generally, in the right context. Affordable action means are incrementally constructed, often unconsciously; when they stabilize, they become social objects. As Saadi Lahlou already said, "People experience difficulties in interpreting new things for which there is no social representation" (Lahlou, 2011).

The use of social representations in multi-agent communications is the most sophisticated model of interaction that I previously denoted **communication by mutual understanding** (Boy, 2002). However, we should not forget that there are two other possible models of interaction that are **supervision** and **mediation**. In the former, a knowledgeable agent supervises the interaction between the other agents. This happens when interacting agents do not know each other, that is, when they do not have appropriate social representations. In human–machine interaction, the supervisor could take the form of an instruction manual, a context-sensitive help, or an expert person. In the latter model, agents may not have social representations of the other agents, but they can communicate through a meditating space. Mediation could be done via other agents usually called facilitators, diplomats, or lobbyist. User-friendly interfaces, and the desktop metaphor in particular, that were developed during the eighties, are typical examples of such meditating spaces. These systemic interaction models form a continuum (Boy, 2009), and they are needed to support job and role analyses.

Emergence and evolution

Emergence is a matter of complexity. It is system's activity that enables the emergence of new functions and structures (Figure 4.6). A good way to discover these emerging functions and structures at design time is to run HITLSs. Note that HITLS can be used at any time during the life cycle of a system to verify possible emergence of needed functions and structures. For example, gesture interaction with computers was a well-recognized invention in HCI for a long time (Sato, Poupyrev & Harrison, 2012), but it took a clever integration with the iPhone to create emerging functions such as the ones we use every day on any smartphones and tablets. The emergence of a new color by an artist is another example. Indeed, by combining two colors, an artist makes emerge a new one. We call this creativity. It is actually integration. How do you measure the sustainability of such emerging functions and structures? You need to construct indicators from experience, expertise, and educated common sense.

Evolution is a matter of adaptation. Complex systems adapt both intrinsically as a system of systems and extrinsically as a component of a system of systems. I would like to claim that people adapt to almost anything that does not violate the laws of physics and physiology!

However, machines need to be ingeniously programmed to adapt to both internal and external disturbances. At this point, we need to see systemic adaptation not only short term, but also longer term. Autopilots, flight management systems, and other automata that are fairly well used today are based on short-term adaptation. Longer-term adaptation of technology is more complex and involves more qualitative research. First, it involves mastering emergent functions and structures discovery processes. Second, it involves mastering articulation of various possible activities of systems of systems (e.g., improve understanding of shared SA, decision-making, and action taking at various levels of granularity of a system of systems). Third, it involves defining local and global metrics at various levels of granularity of a system of systems. Finally, using systemic interaction models will help in analysis, design, and evaluation of system-of-systems activity.

Usability, usefulness, and user experience

Usability and usefulness need to be tested in a wide range of situations. Usability labs have shown their relevance and limitations (Bastien, 2010). No usability and usefulness assessment cannot be possible without accessing user experience (now known under the acronym UX), which is nothing else than user's activity (using the distinction between task and activity).

Nielsen's metrics are very useful. They include the following (Nielsen, 1993): learnability, retention, efficiency, human error handling, and subjective satisfaction. Even if usability is the buzz word for assessing a user interface (UI), the term "usefulness" is broader and includes usability. MacDonald (2012) showed through quantitative and qualitative analyses that "usefulness of a system is shaped by the context in which it is used, that usability is a major element of usefulness, that usefulness has both pragmatic (e.g., usability, simplicity) and hedonic (e.g., aesthetics, pleasurable interactions) attributes, and that usefulness plays a pivotal role in defining users' overall evaluation of a system (i.e., its goodness)."

Usability evolved over the years, eventually leading to the term "user experience," which is about all aspects of human–machine interaction (i.e., systems, the way we defined them in this book) including technological products, organizations (e.g., company, governmental agencies, small groups, communities), and people (e.g., services). HCD practice consists in putting people at the center (i.e., meeting exact needs of these people). Make sure that people can use technology without too much training; this depends on the complexity of the system, of course. Technology should be a pleasurable support. It should be simple and elegant.

Be careful when you decide to satisfy user needs. When they express a need, it is usually based on the technology they have right now, not the technology of the future. You should go into what they have to do ultimately (i.e., the goal of their job). Once you understand enough what is needed, best is to show them a mockup or a prototype that they can evaluate – the architect approach!

Capturing and using UX is not only a question of psychology, HCI, or HFE, it is also a matter of engineering and marketing. UX is not just making a UI, it is becoming aware of people's needs and integrate these needs into the design of the overall system. From that point of view, UX and usability are different concepts. Usability was developed to qualify and quantify UIs.

Situation awareness

There are several models of SA. Mica Endsley's SA model is by far the most influential. It is presented on Figure 9.8.

Endsley's model is defined as a sequence of three cognitive processes (or functions):

1. **Perception** of the elements in the environment, which define the current situation
2. **Comprehension** of the current situation (i.e., a comprehensive holistic pattern or tactical situation – operationally relevant terms in support of rapid decision-making and action)

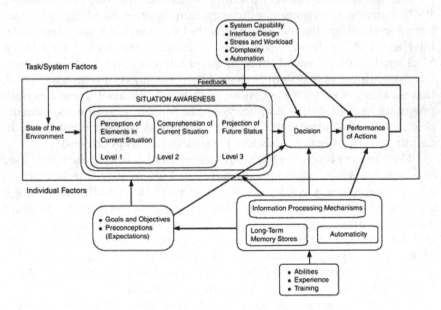

Figure 9.8 Generic SA-centered decision-making model (Endsley, 1995).

3. **Projection** of future status (i.e., attempt to predict the evolution of the tactical situation – support of short-term planning and option evaluation when time permits).

We already discussed the concept of situation in Chapter 3, as: real, available, perceived, expected, and desired. We could add two more types of situation, such as comprehensive situation and projected situation. Note that comprehensive situation is more elaborated than perceived situation and projected situation may become expected and/or desired. These notions are specifically important in activity analysis. They should be considered to construct appropriate SA metrics (see Chapter 2 to go from meaningful metrics to measurement models to observable variables).

In multi-agent environments, SA should be thought in terms of systemic interaction models (see Chapter 4: supervision, mediation, and cooperation) and multi-agent coordination using a set of rules by either supervisors, mediators, or agents themselves.

Key SA models were recently analyzed, divided into individual SA, team SA, and systems SA categories (Stanton et al., 2017). This review shows the importance of considering all three types of models and achieving a match between them and the problem at hand.

Situated actions and context issues

Lucy Suchman coined the term "Situated Action" to denote a theory of human activity (Suchman, 1987). She claimed that nothing can be understood without first understanding its context. I claim that this is an approach to HSI that is centered on **context and culture**, like distributed cognition (Hutchins, 1995). It requires contextual inquiry (Holtzblatt & Beyer, 1993), by always observing, inquiring, and documenting.

Situated actions are related to context. Environment provides context to people's actions (e.g., an obstacle on the road forces you to go around it – in other words, since the obstacle is potential danger for you, you avoid it by going around it). An **affordance** is a potentiality (e.g., what is an affordance of an obstacle? It is the suggestion to avoid it by going around it). James Gibson coined the term "affordances" to denote what the environment offers an individual (Gibson, 1979). In design, it is used to denote the relationship between an object and an individual. Usually, affordances are discovered by experience. This is the reason why we need to develop HITLS to discover such affordances.

It is possible to create useful affordances by developing **virtual reality** images that purposefully increase SA in context. Virtual reality is a technology that superimposes a computer-generated image on a user's view of the real world, thus providing a composite view in the right context.

Augmented reality is extension of virtual reality. It consists in adding useful information on top of real scenes to increase figurative tangibility.

Cooperation and coordination

People work together using various kinds of support to communicate, cooperate, and coordinate their activities. Direct communication, paper documentation, telephone, electronic mail, Internet, Intranets, mobile computing, and desktop conferencing are examples of such supports. People communicate, share information, and coordinate activities synchronously or asynchronously, in the same place or remotely.

Douglas Engelbart is certainly the father of the technology that supports collaborative processes today. He invented the mouse and was one of the engineers who worked on the ARPANET project in the 1960s. He was among the first researchers who developed hypertext technology and computer networks to augment the intellectual capacities of people. The term "Computer-Supported Cooperative Work" (CSCW) was coined in 1984 by Paul Cashman and Irene Grief to describe a multidisciplinary approach focused on how people work and how technology could support them.

Summary

Evaluation processes and metrics (Figure 9.9) need to be defined for each system design and development. They require modeling and simulation tools; methods essentially based on HCD (within the scope of this book); metrics that are based on models, expertise, and experience; and this book proposes a TOP model systematic approach for the definition of evaluation processes and metrics.

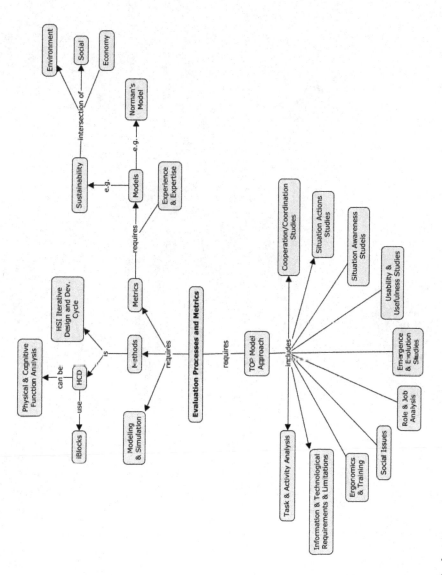

Figure 9.9 Evaluation processes and metrics.

Conclusion

> You cannot hope to build a better world without improving the individuals. To that end, each of us must work for his own improvement and, at the same time, share a general responsibility for all humanity, our particular duty being to aid those to whom we think we can be most useful.

> **Marie Curie**
> *Physicist and Chemist, Nobel Prize*

This book is about current evolution of Human–Systems Integration (HSI), where computers and software are invading all parts of our lives. Does this evolution improve our lives? Who is in charge of designing our sociotechnical world? Who should be? What kinds of methods and tools should we develop to improve us? These questions are difficult to answer without an appropriate philosophical and technical language. I hope this book will contribute to this endeavor.

The design/operations distinction was introduced in Chapter 4 to show the importance of the human element in the design and development of engineering systems. We have seen that human-in-the-loop modeling and simulation (HITLMS) offers crucial support to system design and development. Does it improve us, as Marie Curie suggested? The answer is Yes, conditionally to considering tangibility seriously. Therefore, we should go a step beyond and further explore how people should be considered in design. Should they be directly involved? Should we consider them as "persona" (i.e., generic humans having specific qualities)? Should we develop software models and simulations of people? Should we have human indicators that enable measuring human-centeredness of systems being developed? Should we resurrect what people used for centuries that is educated common sense?

As you can see, I have more questions than answers for you to read, but I would like you to think and try to find your own answers to these questions. We need to start a discussion all over the planet Earth on these questions. Human-centered design (HCD) is currently a disciple under development, where everybody should participate. It is your turn. Please don't hesitate to send me your thought. Exchange with others on the topic. Modeling people using technology (i.e., usages) is key, but we should be careful with the level of details involved. Human modeling led to very sophisticated models that were difficult to use and maintain (e.g., MIDAS, Corker and Smith, 1993). It is often much more effective to use models with higher level of granularity that provide guidance in HCD based on HITLMS instead of implementing a full simulation that includes software-based human models. Note that some models with very low level of granularity could be very helpful for fixing detail problems.

For the last half century, we designed and developed increasingly automated machines. This **automation** contributed to the rigidification of various tasks that human beings have to perform. Automation mainly consists in implementing (rigid) software procedures. Today, we can make technology more flexible by associating appropriate artificial intelligence (AI) algorithms and HCD. It is time to have a discussion on what AI really is and is for. During the 1980s, AI developed as a **goal-driven** discipline. Symbolic approaches were dominant (i.e., symbolic reasoning and symbolic machine learning were prominent). Supporting software languages were Lisp and Prolog. Nowadays, AI resurrected as an **event-driven** discipline. Data science approaches have become dominant (i.e., big data analysis and data-intensive machine learning are prominent). Supporting software languages are TensorFlow Python and Random Forest). We have shifted from qualitative to quantitative approaches to AI. Machine learning and deep learning in particular are now mostly thought as neural network applications. From a philosophical point of view, such data science approaches tend to lead to short-term predictions (i.e., reactive and therefore event-driven), as symbolic AI approaches force the development of ontologies leading to the investigation of longer-term possible futures (i.e., intentional and therefore goal-driven). This distinction needs to be seriously considered because it has direct repercussions on governance.

Instead of trying to predict the future from causal events, it is often better to anticipate **possible futures** – this is where creativity comes to play – and test them based Human-In-The-Loop Simulations. Anticipation is often performed by science fiction novelists, and we need to wait for other people to make their novels or plays tangible. For example, the word "robot" was introduced in English from the phrase *Rossum's Universal Robots*, translated from the Czech language ("Rossumovi Univerzální Roboti" or R.U.R.), a science fiction play written by Karel Čapek in 1920 (Roberts, 2006). It took several decades before tangible robots were developed. George Devol invented the first digitally operated and programmable robot, named the Unimate, in 1954 (Nof, 1999). In contrast, it is interesting to note that Clement Ader, the inventor of the first flying machine, Eole, which flew in 1890 in the vicinity of Paris, also coined the word "avion," which means aircraft in French. In any case, we always need to coin terms (or words) to denote concepts that were not defined before. These terms are either anticipated (e.g., robot) or real immediate necessity (e.g., *avion*). This is the reason why we need to develop ontologies that support design processes.

Sustainability, as an intersection of social, environmental, and economic factors, should drive the discussion. Sustainability is one of the main components of system tangibility (i.e., the others are complexity, maturity, stability, and flexibility). Do we want sustainability in the short term or the longer term? Probably both, but we need to be clear on what we are doing. In any case, we are entering a new era – in the **search for operations tangibility** – which requires finding methods and tools that enable moving **from rigid automation to flexible autonomy**. AI and HSI are at the heart of this evolution.

HSI can be seen as an intersection of human factors and ergonomics (HFE), systems engineering (SE), and computer science (CS) (Figure 1). Within the context of this book, HFE can be seen as task and activity analysis, human and organizational performance evaluation methods, and metrics; SE can be seen as systems of systems, agile development, and design thinking, as well as system thinking, model-based SE; and CS can be seen as human–computer interaction, AI, visualization techniques, as well as modeling and simulation.

A contemporary question is: do we need to develop a new discipline of HSI? As already explained, HSI is an intersection of other disciplines such as HFE, SE, and CS, together with environmental science and economics.

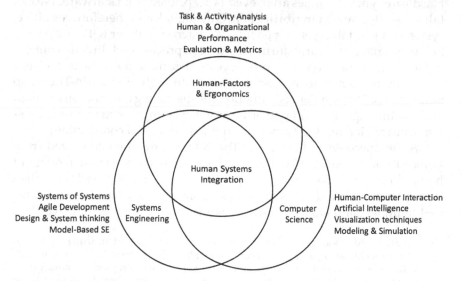

Figure 1 HSI as an intersection of HFE, SE, and CS.

Engineers and top managers should be able to understand the role of humans and organizations in the design of complex systems. Therefore, **HSI should be taught in engineering curricula**. It is also clear that engineering design is a matter of collaborative work between specialists in HFE, SE, and CS, as well as domain experts (e.g., aeronautics, chemical engineering, telecommunication, environmental science, economics).

Why tangibility is suddenly becoming a crucial concept? This is because we inverted the way we design and develop technological systems, using digital technology. We used to start from tangible parts and assemble them to make more complex objects or machines. This was a matter of hardware and structures. We can talk about **physical tangibility**. Software came into play at the end of the 20th century to the point that machines became increasingly automated and interactive, introducing functionalities into machines. Machines that were operated physically were incrementally replaced by systems that are now managed cognitively. We moved from doing to thinking. However, if these automated machines led to safer, more efficient, and comfortable operations, they also introduced rigidity in abnormal, unanticipated, and emergency situations. This the reason why there is a need for investigating the concept of flexible operations from three viewpoints: technology, organizations, and people[1] (the TOP model).

It is interesting to notice that for the last two decades engineering design starts on a computer (using a PowerPoint presentation, for example) and continues to be developed as a virtual model, mockup, and simulation involving people in the loop. This is a matter of software and functions. Hardware typically comes after, even is 3D printed from software. We can talk about **figurative tangibility** of highly digital, and therefore cognitive, systems. A good thing is that people can interact with (or within) systems being designed very early during the design process and directly contribute to improve them along their life cycles. It is, therefore, time to consider autonomy of both humans and machines through Human-In-The-Loop Simulations. This implies coordination means among autonomous human and machine agents. In other words, we need to improve mastering of systemic interaction models (supervision, mediation, and cooperation).

As an anecdote, at the turn of the 20th century, people moved from horses to automobiles. Coaches need to understand mental models of horses in order to manage horses pulling stagecoaches. Coach's job then became automobile driver's job. Drivers, called chauffeurs at that time, had to understand engine and chassis mechanics in order to drive their

---

[1] In the TOP model, the term "people" denotes anybody involved in using developed technology within an organization. It could be replaced by the term "users" and include what is commonly called "end users" but also engineering designers, manufacturers, maintainers, trainers, legislators (in charge of technology certification, for example), and dismantlers.

bosses seating in the back of the automobile. The main disruption was the shift from managing horses to controlling automobiles. After one century, people were able to drive cars without understanding mechanics – technological maturity was reached. Today, in the beginning of the 21st century, automobiles are becoming more autonomous (i.e., they include more intelligent software that provides them with a kind of mental model, like horses have!). We now need to figure out what kind of mental model awareness drivers should have of semiautonomous automobile software… The main disruption was the shift from controlling automobiles to managing autonomous transportation means.

To sum up, I hope that, in addition to providing an approach to HSI based on the search for tangibility using virtual HCD, this book will improve knowledge on and applicability of the **autonomy** concept not only regarding new technologies being currently developed but also and foremost regarding people and organizations that are or will be using them (i.e., autonomous machines but also provision of more autonomy to people and organizations), with the central issue of coordination of autonomous (or semiautonomous) agents. Many situations require human presence to articulate operations in terms of trust, cognitive, and emotional comfort of end users. More specifically, this book focuses on **HCD for operations tangibility** of increasingly autonomous complex systems. Examples are management of vehicles with various levels of autonomy, digitalization of remote complex operations, management of a fleet of robots, training of personnel using real-time simulation systems, or use of digital twins for failure detection and recovery as well as incremental integration of experience feedback knowledge. Details on these kinds of applications will be the content of another book.

References

Aamodt, A. & Plaza, E. (1994). Case-based reasoning: Foundational issues, methodological variations, and system approaches. *Artificial Intelligence Communications*, 7(1), pp. 39–52.

Akao, Y. (1997). QFD: Past, present, and future. *Quality Progress*, 97(2), pp. 1–12.

Alter, N. (1993). La lassitude de l'acteur de l'innovation. *Sociologie Du Travail*, 35(4), pp. 447–468.

Amalberti, R., de Montmollin, M. & Theureau, J. (1991). *Modèles et analyses du travail (Work Models and Analyses)*. Mandaga, Persée, Paris, France.

Ambrosino, J. (2018). Apports de l'hybridation de méthodes de créativité pour l'émergence de projets collaboratifs d'innovation dans les pôles et clusters. *Ph.D. Dissertation*. University of Bordeaux & ESTIA Institute of Technology, France.

Ashby, W.R. (1952). *Design for a Brain*. Wiley, Oxford.

Athènes, S., Averty, P., Puechmorel, Daniel Delahaye, S. & Collet, C. (2002). ATC complexity and controller workload: Trying to bridge the gap. In *Proceedings of HCI-Aero 2002*, G.A. Boy, J. Hansman, Eds. AAAI Press, Cambridge, MA, USA.

Atkinson, D.J. & Clark, M.H. (2013). Autonomous Agents and Human Interpersonal Trust: Can We Engineer a Human-Machine Social Interface for Trust? *Proceedings of 2013 AAAI Spring Symposium Series on Trust and Autonomous Systems*. Stanford University, Stanford CA, USA.

Atkinson, D.J., Friedland, P. & Lyons, J.B. (2012). Human-Machine Trust for Robust Autonomous Systems. *Proceedings of the 7th ACM/IEEE International Conference on Human-Robot Interaction (HRI 2012)*, Boston, MA, USA.

Bastien, C. (2010). Usability testing: A review of some methodological and technical aspects of the method. *International Journal of Medical Informatics*, 79, pp. e18–e23.

Beck, K. et al. (2001). Agile Manifesto. www.agilemanifesto.org. (retrieved on January 26, 2015).

Bellet, T., Boverie, S., Boy, G. & Hoc, J.M. (2010). De l'interaction à la coopération homme-machine: vers le copilotage automobile. In Interaction homme-machine dans les transports, personnalisation, assistance et informations du voyageur, C. Kolski, Ed. Hermes Science Publications, Paris, pp. 151–177.

Bergson, H. (1907). *L'évolution créatrice* (86th ed.). Presses Universitaires de France, Paris.

Bertelsen, O.W. & Bødker, S. (2003). Activity theory. In *HCI Models, Theories, and Frameworks: Toward a Multidisciplinary Science*, J. M. Carroll, Ed. Morgan Kaufmann, San Francisco, pp. 291–324.

Boehm-Davis, D.A., Durso, F.T. & Lee, J.D. (Eds.). (2015). *Handbook of Human Systems Integration*. American Psychology Association, Washington, DC. ISBN 978-1-4338-1828-8.

Boulnois, S.L.P. (2018). Human-Centered Design of a 3D-Augmented Strategic Weather Management System for the Aviation Community. *Doctoral Dissertation*. Florida Institute of Technology, USA.

Boy, G.A. (2017). Human-centered design of complex systems: An experience-based approach. *Design Science Journal*, 3. Cambridge University Press, UK, pp. 1–23.

Boy, G.A. & Brachet, G. (2010). Risk Taking. Dossier of the Air and Space Academy, Toulouse, France, ISBN 2-913331-47-5.

Boy, G.A. & Narkevicius, J. (2013). Unifying human centered design and systems engineering for human systems integration. In *Complex Systems Design and Management*, M. Aiguier, F. Boulanger, D. Krob & C. Marchal, Eds. Springer, Paris, UK, 2014. ISBN 978-3-319-02811-8.

Boy, G.A. & Schmitt, K.A. (2012). Design for safety: A cognitive engineering approach to the control and management of nuclear power plants. *Annals of Nuclear Energy*, 52, pp. 125–136. Elsevier.

Boy, G.A. & Tessier, C. (1985). Cockpit Analysis and Assessment by the MESSAGE Methodology. *Proceedings of the 2nd IFAC/IFIP/IFORS/IEA Conf. on Analysis, Design and Evaluation of Man-Machine Systems*, Villa-Ponti, Italy, September 10–12. Pergamon Press, Oxford, pp. 73–79.

Boy, G.A. (1981). A Numerical Method for Convective and Diffusive Process Simulation, Application to Respiratory Gas Transfer. *Proceedings of the 2nd World Congress of Chemical Engineering*, Montreal, Canada.

Boy, G.A. (1983). Le Système MESSAGE: Un Premier Pas vers l'Analyse Assistée par Ordinateur des Interactions Homme-Machine [A first step toward computer-assisted analysis of human-machine interactions]. *Le Travail Humain Journal*, 46(2), pp. 271–286.

Boy, G.A. (1991a). *Intelligent Assistant System*. Published by Academic Press, London. ISBN 0121212459.

Boy, G.A. (1991b). Advanced Interaction Media as a Component of Everyday Life for the Coming Generation. Proceedings of the World Marketing Congress, Japan Management Association, Tokyo, Japan.

Boy, G.A. (1996). The Group Elicitation Method: An Introduction. *Proceedings of EKAW'96, Lecture Notes in Computer Science Series*, Springer Verlag, Berlin.

Boy, G.A. (1997a). The group elicitation method for participatory design and usability testing. *Interactions*, 4(2), March-April, ACM, New York, pp. 27–33.

Boy, G.A. (1997b). Active Design Documents. *Conference Proceedings of ACM DIS'93 (Designing Interactive Systems)*. ACM Digital Library, New York, USA.

Boy, G.A. (1998). *Cognitive Function Analysis*. Praeger/Ablex, Stamford, CT, USA; ISBN 9781567503777.

Boy, G.A. (2002). Theories of human cognition: To better understand the co-adaptation of people and technology. In *Knowledge Management, Organizational Intelligence and Learning, and Complexity*, L. Douglas Kiel, Ed. Encyclopedia of Life Support Systems (EOLSS), Developed under the Auspices of the UNESCO, Eolss Publishers, Oxford, UK.

Boy, G.A. (2003). *L'Ingénierie Cognitive: Interaction Homme-Machine et Cognition (The French Handbook of Cognitive Engineering)*. Lavoisier, Paris, Hermes Sciences.

Boy, G.A. (2005). Knowledge Management for Product Maturity. *Proceedings of the International Conference on Knowledge Capture (K-Cap'05)*. Banff, Canada. October.

Boy, G.A. (2009). The Orchestra: A Conceptual Model for Function Allocation and Scenario-Based Engineering in Multi-Agent Safety-Critical Systems. *Proceedings of the European Conference on Cognitive Ergonomics*, Otaniemi, Helsinki area, Finland; 30 September-2 October.

Boy, G.A. (2010). An Architecture for Socio-Cultural Modeling. Proceedings of the Third International Conference on Applied Human Factors and Engineering, Miami, Florida; 17–20 July 2010. Also in D. Schmorrow & D. Nicholson (Eds.), Advances in Cross-Cultural Decision Making, CRC Press 2010, pp. 250–259; Print ISBN: 978-1-4398-3495-4; eBook ISBN: 978-1-4398-3496-1; Doi: 10.1201/EBK1439834954-c26.

Boy, G.A. (2011). (Ed.) *Handbook of Human-Machine Interaction: A Human-Centered Design Approach*. Ashgate, Farnham, UK. ISBN 978-1-1380-7582-5.

Boy, G.A. (2013a). *Orchestrating Human-Centered Design*. Springer, London, UK, ISBN 978-1-4471-4338-3.

Boy, G.A. (2013b). Dealing with the Unexpected in our Complex Socio-Technical World. *Proceedings of the 12th IFAC/IFIP/IFORS/IEA Symposium on Analysis, Design, and Evaluation of Human-Machine Systems*. Las Vegas, Nevada, USA.

Boy, G.A. (2016). *Tangible Interactive Systems*. Springer, Cham, Switzerland, UK, ISBN 978-3-319-30270-6.

Boy, G.A., Jani, G., Manera, A., Memmott, M., Petrovic, B., Rayad, Y., Stephane, A.L. & Suri, N. (2016). Improving collaborative work and project management in a nuclear power plant design team: A human-centered design approach. *Annals of Nuclear Energy*, 94, pp. 555–556. Elsevier, ANE4864.

Boy, G.A., Jorda, M.F., Renun, B., Brieussel, J.M., Caval, E. & Lareng, L. (1980). Comparison of Different Methods of Mean Alveolar PCO2 Measurement during Mechanical Ventilation. *Proceedings Computer in Critical Care and Pulmonary Medicine*, IEEE, Elsevier, Holland. Lund, Sweden.

Cadoz, C., Luciani, A., Villeneuve, J., Kontogeorgakopoulos, A. & Zannos, I. (2014). Tangibility, Presence, Materiality, Reality in Artistic Creation with Digital Technology. *40th International Computer Music Conference / 11th Sound and Music Computing Conference*, National and Kapodistrian University of Athens, Sep, Athens, Greece. pp. 754–761.

Calvary, G., Coutaz, J., Thevenin, D., Limbourg, Q., Bouillon, L. & Vanderdonckt, J. (2003). A unifying reference framework for multi- target user interfaces. *Interacting with Computers*, 15(3), pp. 289–308.

Cano, E.L.L., Moguerza, J.M. & Redchuk, A. (2012). Six Sigma with R: Statistical Engineering for Process Improvement (Use R!). Springer, New York. ISBN 978-1461436515.

Card, S.K., Moran, T.P. & Newell, A. (1983) *The Psychology of Human-Computer Interaction*. Erlbaum, Hillsdale.

Caruso, P., Dumbacher, D. & Grieves, M. (2010). Product Lifecycle Management and the Quest for Sustainable Space Explorations. *AIAA SPACE 2010 Conference & Exposition. Anaheim, CA*.

Castellani, B. & Hafferty, F. (2009). *Sociology and Complexity Science*. Springer, Paris, UK, ISBN 978-3-540-88462-0.

Checkland, P. (1999). *Systems Thinking, Systems Practice.* John Wiley & Sons, New York, NY, USA.

Christensen, C. (1992). Exploring the limits of the technology S-curve. Parts I and II. *Production and Operations Management,* 1(4), pp. 334–366.

Clarke, J.P. & Lauber, J. (2014). *Autonomy Research for Civil Aviation: Toward a New Era of Flight.* The National Academies Press, Washington, DC. 20001. ISBN 978-0-309-38688-3; Doi: 10.17226/18815.

Cohen, S., Bose, S., Guo, Miller, A., DeFrancia, K., Berger, O., Filiatraut, B., Loman, M., Qiu, W. & Zhang, C. (2014). The Growth of Sustainability Metrics. Sustainability Metrics White Paper Series: 1 of 3. Columbia University Earth Institute. (retrieved on August 4, 2019: http://spm.ei.columbia.edu/files/2015/06/SPM_Metrics_WhitePaper_1.pdf).

Conroy, M. (2016). Playing Nice Across Time & Space: Tools, Methods and Tech for Multi-Location Multi-Decadal Teams. *INCOSE International Workshop, HSI Working Group Meeting,* Torrence, California, USA.

Cooper, G. & Herskovits, E. (1992). A Bayesian method for the induction of probabilistic networks from data. *Machine Learning,* 9, pp. 309–347.

Cooper, G.E. & Harper, R.P. (1969). The Use of Pilot Rating in the Evaluation of Aircraft Handling Qualities. AGARD Report 567, NATO, Advisory Group for Aerospace Research and Development.

Cooper, G.E., White, M.D., & Lauber, J.K. (Eds). (1980). Resource Management on the Flightdeck: Proceedings of a NASA/Industry Workshop (NASA CP-2120). Moffett Field, CA: NASA-Ames Research Center.

Corker, K.M. & Smith, B.R. (October, 1993). An Architecture and Model for Cognitive Engineering Simulation Analysis: Application to Advanced Aviation Automation. The Proceedings of the AIAA Computing in Aerospace 9 Conference. October, 1993: San Diego, CA, USA.

Coze, Y., Kawski, N., Kulka, T., Sire, P., Sottocasa, P. & Bloem, J. (Eds.) (2009). *Virtual Concept > Real Profit with Digital Manufacturing and Simulation.* Dassault Systèmes and Sogeti. LINE UP Book & Media, The Netherlands. ISBN 978 90 75414 25 7.

Cumming, G. & Calin-Jageman, R.J. (2017). *Introduction to the New Statistics: Estimation, Open Science, and Beyond.* Routledge, New York.

DAU. (February 19, 2010). *Defense Acquisition Guidebook (DAG).* Defense Acquisition University (DAU)/U.S. Department of Defense (DoD), Ft. Belvoir, VA, USA.

Desberg, P. & Taylor, J. (1986). *Essentials of Task Analysis.* University Press of America, Lanham, MA.

Drury, C.G. (1983). Task analysis methods in industry. *Applied Ergonomics,* 14(1), pp. 19–28.

Dubois, D. & Prade, H. (2011). Possibility theory, probability theory and multiple-valued logics: A clarification. *Annals of Mathematics and Artificial Intelligence,* 32(1–4), pp. 35–66. Kluwer Academic Publishers, Norwell, MA.

Efron, B. (June, 2013). Bayes' theorem in the 21st century. *Science,* 340(6137), pp. 1177–1178. Doi: 10.1126/science.1236536.

Endsley, M. (1995). Measurement of situation awareness in dynamic systems. *Human Factors,* 37, pp. 65–84.

Engelbart, D.C. (1986). Workstation History and the Augmented Knowledge Workshop. *Proceedings of the ACM Conference on the History of Personal Workstations,* Palo Alto, CA, pp. 73–83. (AUGMENT, 101931). Republished as the Augmented Knowledge Workshop in "A History of Personal Workstations," Adele Goldberg (Ed.), ACM Press, New York, 1988, pp. 185–236.

Espinosa, J.A. (2007). Familiarity, complexity, and team performance in geographically distributed software development. *Organization Science*, 18, pp. 613–630.

Fehrm, B. (2018). Bjorn's Corner: Aircraft stability, Part 2. (Retrieved on June 9, 2019). https://leehamnews.com/2018/04/20/bjorns-corner-aircraft-stability-part–2/.

Fisher, D. (1987). Knowledge acquisition via incremental conceptual clustering (PDF). *Machine Learning*, 2(2), pp. 139–172. doi:10.1007/BF00114265.

Fitts, D.J., Sándor, A., Litaker, H.L. & Tillman, B. (2008). Human Factors in Human-Systems Integration. Presentation at HSI meeting of INCOSE International Workshop.

Flood, R.L. (1999). *Rethinking the Fifth Discipline: Learning within the Unknowable.* Routledge, London, UK.

Folds, D. (2015). Systems engineering perspective on human systems integration. In *Systems Thinking: Coping with 21st Century Problems*, J. Boardman, B. Sauser, Eds. Taylor & Francis, Boca Raton, FL, USA.

Foster, R.N. (1986). Working the S-curve: assessing technological threats. *Research Management*, 29(4), pp. 17–20.

Friedland, P.E. & Iwasaki, Y. (June 1985). The concept and implementation of skeletal plans. *Journal of Automated Reasoning*, 1(2), pp. 161–208.

Gaines, B.R. & Boose, J.H. (1988). *Knowledge Acquisition for Knowledge-Based Systems*. Academic Press, Orlando, FL, USA. SBN:0122732510.

Gaines, B.R. (February 2013). Knowledge acquisition: Past, present and future. *International Journal of Human-Computer Studies*, 71(2), pp. 135–156. Doi: 10.1016/j.ijhcs.2012.10.010.

Garcia Belmonte, N. (2016). Engineering Intelligence through Data Visualization at Uber. (retrieved on July 29, 2019: https://eng.uber.com/data-viz-intel/).

Gibbons, S. (2018). Journey Mapping 101. (retrieved on August 4, 2019: www.nngroup.com/articles/journey-mapping-101/).

Gibson, J.J. (1979). *The Ecological Approach to Visual Perception*. Houghton Mifflin, Boston, USA. ISBN 0898599598.

Glaessgen, E.H. & Stargel, D. (2012). The Digital Twin Paradigm for Future NASA and us Air Force Vehicles. AAIA 53rd Structures, Structural Dynamics, and Materials Conference. Honolulu, Hawaii.

Goodrich, M. & Olsen, D. (2003). Seven Principles of Efficient Human Robot Interaction. Proceedings of IEEE International Conference on Systems, Man and Cybernetics, Washington, DC, pp. 3943–3948.

Greenbaum, J. & Kyng, M. (Eds). (1991). *Design at Work: Cooperative Design of Computer Systems*. Lawrence Erlbaum Associates, Publishers, Hillsdale, NJ.

Gregory, J. (2003). Scandinavian approaches to participatory design. *International Journal of Engineering*, 19(1), pp. 62–74, TEMPUS Publications.

Grieves, M. (2016). Origins of the Digital Twin. Concept Working Paper. August. Florida Institute of Technology / NASA, Doi: 10.13140/RG.2.2.26367.61609.

Griffin, G. (2010). Crew-Ground Integration in Piloted Space Programs. *Keynote speech at HCI-Aero Conference, Cape Canaveral*, Florida, USA.

Gruber, T.R. (1993a). A translation approach to portable ontology specification. *Knowledge Acquisition*, 5, pp. 199–220.

Gruber, T.R. (1993b). Toward principles for the design of ontologies used for knowledge sharing. *International Journal Human-Computer Studies*, 43, pp. 907–928. Available as Technical Report KSL 93–04, Knowledge Systems Laboratory, Stanford University.

Grudin, J. (1994). Computer-supported cooperative work: History and focus. *Computer*, 27(5), pp. 19–26. Doi: 10.1109/2.291294.

Guidance Notes on the Application of Ergonomics to Marine Systems. (2013–2018). American Bureau of Shipping Incorporated by Act of Legislature of the State of New York 1862. American Bureau of Shipping. ABS Plaza, 16855 Northchase Drive, Houston, TX 77060 USA.

Hackos, J.T. & Redish, J.C. (1998). *User and Task Analysis for Interface Design*. John Wiley & Sons, Inc, New York.

Hansman, R.J., Pritchett, A. & Midkiff, A. (1995). Party Line Information Use Studies and Implications for ATC Datalink Communications. In Proceedings of the Fifth International Conference On Human-Machine Interaction and Artificial Intelligence in Aerospace, G.A. Boy, Toulouse, France, September 27–29.

Hart, S. & Staveland, L. (1988). Development of NASA-TLX (Task Load Index): Results of empirical and theoretical research. In *Human Mental Workload*, P. Hancock, N. Meshkati, Eds. Elsevier Science, North Holland.

Hayakawa, S.I. (1990). *Language in Thought and Action*. (5th ed.). Harcourt, Brace & Jovanovich, New York.

Helmreich, R.L., Merritt, A.C & Wilhelm, J.A. (1999). The evolution of Crew Resource Management training in commercial aviation. *International Journal of Aviation Psychology*, 9(1), pp. 19–32.

Hofweber, T. (2005). A puzzle about ontology. *Noûs*, 39(2), pp. 256–283. (retrieved on July 31, 2019: https://philpapers.org/rec/HOFAPA–2).

Hofweber, T. (2018). Logic and ontology. In E. N. Zalta (Ed.) The Stanford Encyclopedia of Philosophy (Summer 2018 Edition). Stanford, CA: Metaphysics Research Lab, Stanford University. Available online at https://plato.stanford.edu/archives/sum2018/entries/logic-ontology.

Hollnagel, E. & Woods, D.D. (1983). Cognitive systems engineering: New wine in new bottles. *International Journal of Man-Machine Studies*, 18, pp. 583–600.

Hollnagel, E. (1993). *Human Reliability Analysis: Context and Control*. Academic Press, London, UK.

Hollnagel, E. (1998). *Cognitive Reliability and Error Analysis Method*. Elsevier, London, UK.

Hollnagel, E. (2014). Functional Resonance Assessment Method. (retrieved from Internet on 9 July 2019: www.spek.fi/loader.aspx?id=2fe8ba37-e4da-4fce-9dfa-3ca6915cd603).

Holtzblatt, K. & Beyer, H. (1993). Making customer-centered design work for teams. *Communications of the ACM*, 36(10), 92–103. October.

Horridge, M., Knublauch, H., Rector, A., Stevens, R. & Wroe, C. (2011). *A Practical Guide to Building OWL Ontologies Using the Protégé-OWL Plugin and CO-ODE Tools Edition 1.3*. The University of Manchester, Manchester, UK.

http://tryqa.com/what-is-v-model-advantages-disadvantages-and-when-to-use-it?"

Hubault, F. (1995). What is the purpose of analyzing activity in ergonomics? [A quoi sert l'analyse de l'activité en ergonomie?] White Paper. Département Ergonomie et Écologie Humaine - Paris 1.

Hutchins, E. (1995) How a cockpit remembers its speeds. *Cognitive Science,* 19, pp. 265–288.

Jaynes, E.T. (1988). How does the brain do plausible reasoning? In *Science and Engineering -- Maximum-Entropy and Bayesian Methods,* G.J. Erickson, C.R. Smith, Eds. Kluwer, Dordrecht.

Jonassen, D.H., Hannum, W.H. & Tessmer, M. (1999). *Task Analysis Methods for Instructional Design.* Lawrence Erlbaum, Mahwah, NJ, USA.

Kanki, B. (2018). Human factors research methods and tools. In *Space Safety and Human Performance,* T. Sgobba, B. G. Kanki, J.-F. Clervoy, and G. Sandal, Eds. Butterworth-Heinemann, Cambridge, MA, pp. 239–271. Doi: 10.1016/B978-0-08-101869-9.00007-8.

Kaptelinin, V. & Nardi, B.A. (2006). *Acting with Technology: Activity Theory and Interaction Design.* MIT Press, Cambridge.

Keim, D., Andrienko, G., Fekete, J.D., Carsten Görg, C., Kohlhammer, J. & Melançon, G. (2008). Visual analytics: Definition, process and challenges. In *Information Visualization - Human-Centered Issues and Perspectives,* A. Kerren, J.T. Stasko, J.-D. Fekete, C. North, Eds. Springer, LNCS, pp. 154–175.

Kienberger, S. & Tiede, D. (2008). ArcGIS explorer review. *GEO Informatics,* 11, pp. 42–47.

Kirwan, B. & Ainsworth, L.K. (Ed.) (1983). *A Guide to Task Analysis.* Taylor and Francis, London.

Kozulin, A. (1986). The concept of activity in Soviet psychology: Vygotsky, his disciples and critics. *American Psychologist,* 41(3): 264–274. Doi: 10.1037/0003–066X.41.3.264.

Kruchten, N. (2018). Data Visualization for Artificial Intelligence, and Vice Versa. (retrieved on July 30, 2019: https://medium.com/@plotlygraphs/data-visualization-for-artificial-intelligence-and-vice-versa-a38869065d88).

Lahlou, S. (2004). Cognitive Attractors and Social Representations. *Proceedings of the 7th International Conference on Social Representations.* Guadalrara, Mexico, 10–14 Sept.

Lahlou, S. (2011). Socio-cognitive issues in human-centered design for the real world. In *Handbook of Human-Machine Interaction,* G.A. Boy, Ed. Ashgate, UK.

Laprie, J.C. & Kanoun, K. (1996). Software reliability and system reliability. In *Handbook of Software Reliability Engineering,* M.R. Lyu, Ed. McGraw-Hill, New York, pp. 27–69.

Laurain, T., Boy, G.A. & Stephane, A.L. (2015). Design of an On-Board 3D Weather Situation Awareness System. *Proceedings 19th Triennial Congress of the International Ergonomics Association,* Melbourne, Australia, 9–14 August.

Lee, J.D. & See, K.A. (2004). Trust in automation: Designing for appropriate reliance. *Human Factors,* 46(1), pp. 50–80.

Leplat, J. (2008). *Repères pour l'analyse de l'activité en ergonomie. Le Travail Humain.* Presses Universitaires de France, Paris, France.

Llewellyn, D.J. (2003). The Psychology of Risk Taking Behavior. *PhD Thesis,* The University of Strathclyde. U.K.

Lorenz, E.N. (1963). Deterministic nonperiodic flow. *Journal of the Atmospheric Sciences,* 20(2), pp. 130–141. Doi: 10.1175/1520-0469(1963)020<0130:dnf>2.0.co;2.

Low, T. (2016). Black swan: the impossible bird. *Australian Geographic.* Rertieved on July 17, 2019: www.australiangeographic.com.au/blogs/wild-journey/2016/07/black-swan-the-impossible-bird/.

Lyu, M.R. (1995). *Handbook of Software Reliability Engineering.* McGraw-Hill Publishing, New York, 1995, ISBN 0-07-039400-8.

MacDonald, C.M. (2012). Understanding Usefulness in Human-Computer Interaction to Enhance User Experience Evaluation. *Ph.D. Dissertation,* Information Studies -- Drexel University, USA.

Macfarlane, A. (1977). Historical anthropology (Frazer lecture). *Cambridge Anthropology,* 3(3): http://www.alanmacfarlane.com/TEXTS/frazerlecture. pdf.

Madni, A.M., Madni, C.C. & Lucero, S.D. (2019). Leveraging digital twin technology in model-based systems engineering. *Systems,* 7(1), p. 7. doi: 10.3390/systems7010007.

Mann, D. (1999). Using S-Curves and Trends of Evolution in R&D Strategy Planning. *I Mech E seminar on FutureHeat Pump and Refrigeration Technologies,* held in London on 20 April (retrieved on 27 August 2019: https://triz-journal. com/using-s-curves-trends-evolution-rd-strategy-planning/).

Mantovani, G. (1996). Social context in HCI: A new framework for mental models, cooperation, and communication. *Cognitive Science,* 20, pp. 237–269.

Markie, P. (2013). Rationalism vs. Empiricism. Stanford Encyclopedia of Philosophy. First published August 19, 2004. http://plato.stanford.edu/entries/rationalism-empiricism/ (Retrieved on September 3, 2015.

Mathur, P. (2017). *Technological Forms and Ecological Communication: A Theoretical Heuristic.* Lexington Books, Lanham, Boulder, New York & London, p. 120.

Mayer, S., Lischke, L., Wo´zniak, P.W. & Henze, N. (2018). Evaluating the Disruptiveness of Mobile Interactions: A Mixed-Method Approach. *Proceedings of CHI 2018,* April 21–26, 2018, Montreal, QC, Canada. ACM Digital Library, New York, USA.

Milson, A.J. & Alibrandi, M. (2008). Digital geography: Geospatial technologies in the social studies classroom. In *International Social Studies Forum: The Series.* R.A. Diem, Ed. University of Texas, Jeff Passe, The College of New Jersey, San Antonio.

Minsky, M. (1986). *The Society of Mind.* Simon & Schuster, New York.

Mitchell, M. (2009). *Complexity: A Guided Tour.* Oxford University Press, New York.

Morin, E. (1990). *Introduction à la Pensée Complexe.* ESF, Paris.

Musen, M.A. (1992). Dimensions of knowledge sharing and reuse. *Computers and Biomedical Research,* 25, pp. 435–467.

Musen, M.A. (2015). The protégé project: A look back and a look forward. *AI Matters,* 1(4), pp. 4–12. Doi: 10.1145/2757001.2757003.

Musen, M.A., Tu, S.W., Das, A.K. & Shahar, Y. (1996). EON: A component-based approach to automation of protocol-directed therapy. *Journal of the American Medical Informatics Association,* 3(6), pp. 367–388. Nov/Dec.

Narayanan, S. & Kidambi, P. (2011). Interactive simulations: History, features and trends. In *Human-In-The-Loop Simulations,* S. Narayanan, P. Kidambi, Eds. Springer-Verlag, London, UK. Doi: 10.1007/978-0-85729-6-1.

Nardi, B.A. (1995). *Context and Consciousness: Activity Theory and Human-Computer Interaction.* MIT Press, Cambridge. ISBN 978-0-262-14058-4.

Neal, L. & Mantei, M. (1993). Computer-Supported Meeting Environments. Tutorial Notes. *Conference on Human Factors in Computing Systems, INTERCHI'93.* Amsterdam, The Netherlands.

Newman, W. & Lamming, M. (1995). *Interactive System Design*. Addison-Wesley, Boston.

Nielsen, J. (1986). A virtual protocol model for computer-human interaction. *International Journal of Man-Machine Studies*, 24(1986), pp. 301–312.

Nielsen, J. (1993). *Usability Engineering*. Academic Press, Boston, MA. ISBN 0-12-518405-0.

Nof, S.Y. (1999). *Handbook of Industrial Robotics* (2nd ed.). Wiley, pp. 3–5. ISBN 0-471-17783-0.

Norman, D.A. (1981). Categorization of action slips. *Psychological Review*, 88, pp. 1–15.

Norman, D.A. (1982). Steps Toward a Cognitive Engineering: Design Rules Based on Analyses of Human Error. *Conference Proceedings of the 1982 Conference on Human Factors in Computing Systems*. ACM Digital Library. DOI: 10.1145/800049.801815.

Norman, D.A. (1986). Cognitive engineering. In *User Centered System Design: New Perspectives on Human-Computer Interaction*, D.A. Norman, S.W. Draper, Eds. Lawrence Erlbaum Associates, Hillsdale, NJ.

Norman, D.A. (1992). *Turn Signals Are Facial Expressions of Automobiles*. Addison Wesley, Reading, MA.

Norman, D.A. (2013). *The Design of Everyday Things: Revised and Expanded*. Basic Books, New York. ISBN-13:978-0465050659.

Novak, J.D. & Cañas, A.J. (2006). The Theory Underlying Concept Maps and How to Construct and Use Them. Technical Report IHMC CmapTools 2006-01, Institute for Human and Machine Cognition (IHMC).

Noy, N.F. & McGuinness, D.L. (2019). *Ontology Development 101 – A Guide to Creating Your First Ontology*. Stanford University, Stanford, CA 94305 (retrieved on August 27, 2019; https://protege.stanford.edu/publications/ontology_development/ontology101-noy-mcguinness.html).

NTSB. (2010). Loss of Thrust in Both Engines After Encountering a Flock of Birds and Subsequent Ditching on the Hudson River US Airways Flight 1549 Airbus A320-214, N106US. Weehawken, New Jersey, January 15, 2009. Accident Report NTSB/AAR-10/03 PB2010-910403.

Ochs, M., Nyengaard, J.R., Jung, A., Knudsen, L., Voigt, M., Wahlers, T., Richter, J. & Gundersen, H.J. (2004). The number of alveoli in the human lung. *American Journal of Respiratory Critical Care Medicine*, 169(1), pp. 120–124.

Orfeuil, J.P. & Leriche, Y. (2019). *Piloter le véhicule autonome au service de la ville [Driving the autonomous vehicle in the service of the city]*. Descartes & Cie, Paris, France. ISBN 978-2-84446-332-6.

Palagi, E., Giboin, A., Gandon, F. & Troncy, R. (2018). Un Modèle de Recherche Exploratoire pour l'Évaluation de ses Systèmes Applications. *Proceedings of IHM 2018–30eme Conférence Francophone sur l'Interaction Homme-Machine*, Brest, France, pp. 1–7.

Paternò, F., Mancini, C. and Meniconi, S. (1997). ConcurTaskTrees: A Diagrammatic Notation for Specifying Task Models. In *Proceedings of INTERACT'97*, Sydney, NSW, Australia, pp. 362–369.

Patrick, J. & James, N. (2004). Process tracing of complex cognitive work tasks. *Journal of Occupational and Organizational Psychology*, 77(2), p. 259.

Paulk, M.C., Weber, C.V., Curtis, B. & Chrissis, M.B. (1995). *Capability Maturity Model: The Guidelines for Improving the Software Process*. The SEI Series in Software Engineering, Addison Wesley Professional, Boston.

Pew, R.W. & Mavor, A.S. (2007). *Human-System Integration in the System Development Process: A New Look*. Committee on Human Factors, National Research Council. National Academies Press, Washington, DC. ASIN: B009WZASPQ.

Piascik, R., Vickers, J., Lowry, D., Scotti, S., Stewart, J. & Calomino, A. (2010). Technology Area 12: Materials, Structures, Mechanical Systems, and Manufacturing Road Map, NASA Office of Chief Technologist.

Rasmussen, J. (1983). Skills, rules, and knowledge: Signals, signs and symbols and other distinctions in human performance models. *IEEE Transactions: Systems, Man & Cybernetics*, 3, pp. 257–267. SMC-13. ISSN: 0018-9472.

Rasmussen, J. (1986). *Information Processing and Human-Machine Interaction: An Approach to Cognitive Engineering*. Elsevier Science Inc., New York, NY, USA. ISBN:0444009876.

Reason, J.T. (1990). *Human Error*. Cambridge University Press, Cambridge, UK.

Reason, J. (1997). *Managing the Risks of Organizational Accidents*. Ashgate, Aldershot. ISBN:1840141042.

Rivière, C. (1969). *L'Objet Social*. Marcel Rivière et Companie, Paris, France.

Roberts, A. (2006). *The History of Science Fiction*. Palgrave MacMillan, New York. ISBN 9780333970225.

Rolos, C., Masson, D. & Boy, G.A. (2019). Operations Room Design for the Control of a Fleet of Robots: Ontology-Based Function Identification and Allocation. *Proceedings of INCOSE HSI2019*, Biarritz, France, September 11–13.

Rook, P. (Ed.) (1990). *Software Reliability Handbook*. Centre for Software Reliability, City University, London, UK.

Rosenblueth, A., Wiener, N. & Bigelow, J. (1943). Behavior, purpose, and teleology. *Philosophy of Science*, 10, pp. 18–24.

Rumsey, D. & Williams, M. (2002). Historical Maps in GIS. Past time, Past Place. GIS for history. A. 1 knowels, ESRI Press, pp. 1–18.

Russel, S. & Norvig, P. (2010). *Artificial Intelligence – A Modern Approach* (3rd ed.). Prentice Hall, Boston, USA. ISBN 978-0-13-604259-4.

Sanders, M. & McCormick, L. (1993). *Human Factors in Engineering and Design*. McGraw-Hill, New York, NY, USA.

Sato, M., Poupyrev, I. & Harrison, C. (2012). Touch'e: Enhancing Touch Interaction on Humans, Screens, Liquids, and Everyday Objects. In Proceedings of the SIGCHI Conference on Human Factors in Computing Systems, *CHI'12*, pp. 483–492, ACM, New York, NY, USA.

Schank, R. (1982). *Dynamic Memory: A Theory of Learning in Computers and People*. Cambridge University Press, New York.

Scheff, T.J. (2006). *Goffman Unbound!: A New Paradigm for Social Science*. Paradigm Publishers, Boulder, CO. ISBN 978-1594511967.

Schlundt, D.G. (2011). Reductionism and Complex Systems Science: Implications for Translation Research in the Health and Behavioral Sciences. Behavioral Risk Factor Surveillance System, Center for Disease Control and Prevention (retrieved on 13 July 2019: www.mc.vanderbilt.edu/crc/workshop_files/2012-01-20.pptx.

Schuler, D. & Namioka, A. (Eds.) (1993). *Participatory Design: Principles and Practices*. Lawrence Erlbaum, Hillsdale, NJ.

Schumpeter, J. (1911). *Théorie de l'évolution économique*. Dalloz, Paris.

Schumpeter, J. (1942). *Capitalism, Socialism and Democracy*. Harper & Row, New York, USA.

Schwaber, K. (1997). Scrum development process. In *OOPSLA Business Objects Design and Implementation Workshop Proceedings*, J. Sutherland, D. Patel, C. Casanave, J. Miller, G. Hollowell, Eds. Springer, London, UK.

Sheridan, T.B. (1984). Supervisory Control of Remote Manipulator, Vehicles and Dynamic Processes: Experiment in Command and Display Aiding. Advances in Man-Machine System Research, 1, J.A.I. Press, pp. 49–137.

Simon, H.A. (1996). *The Sciences of the Artificial*. MIT press, Cambridge, MA, USA.

Sonntag, C., Engell, S., Papanikolaou, V., Sarris, N., Watzek, S., Fettweis, G., Sonntag, C., Engell, S. & Cave, J. (2017). Opportunity Report: "Towards Enhanced EU-US ICT Pre-competitive Collaboration". Revised version V1.0.1. (www.picasso-project.eu).

Spérandio, J.C. (1984). *Ergonomics of Mental Work (In French: L'ergonomie du travail mental)*. Masson, Paris, France.

St. Amant, R., Healey, C.G., Riedl, M., Kocherlakota, S., Pegram, D.A. & Torhola, M. (2001). Intelligent Visualization in a Planning Simulation. *Proceedings Intelligent User Interfaces IUI'01)*. Sante Fe, New Mexico. ACM 1-58113-325-1/01/0001.

Stanton, N., Salmon, M., Walker, G.H., Salas, E. & Hancock, P.A. (2017). State-of-science: Situation awareness in individuals, teams and systems. *Ergonomics*, 60(4), pp. 247–258. Doi: 10.1080/00140139.2017.1278796.

Steinfeld, A., Fong, T., Kaber, D., Lewis, M., Sholtz, J., Schultz, A. & Goodrich, M. (2006). Common Metrics for Human-Robot Interaction. *HRI'06 Conference*, March 2–4, Salt Lake City, Utah, USA. ACM Digital Library.

Stix, G. (1994). Aging airways. *Scientific American*, 270(5), pp. 96–104, May.

Suchman, L.A. (1987). *Plans and Situated Actions: The Problem of Human-Machine Communication*. Cambridge University Press, Cambridge, UK.

Sushil, X. (1997). Flexible systems management: An evolving paradigm. *Systems Research and Behavioral Science*, 14(4), pp. 259–275.

Sushil. (1999). *Flexibility in Management. Global Institute of Flexible Systems Management*. Vikas Publishing House, New Delhi.

Sushil. (2000). Systemic flexibility. *Global Journal of Flexible Systems Management*, 1(1), pp. 77–80.

Sutherland, J. (2014). *Scrum: The Art of Doing Twice the Work in Half the Time*. Crown Business, Primento. ISBN-13: 978-0385346450.

Taleb, N.N. (2010). *The Black Swan*. Random House, New York. ISBN-13: 978-0812973815.

Thom, R. (1972). *Stabilité structurelle et morphogénèse: essai d'une théorie générale des modèles*. W. A. Benjamin, Reading, MA.

Thomas, R. & Alaphilippe, D. (1993). *Les Attitudes*. Que sais-je? Presses Universitaires de France, Paris, France.

Tichkiewitch, S. & Brissaud, D. (Eds.) (2013). *Methods and Tools for Co-operative and Integrated Design*. Springer Science & Business Media, London. ISBN 9401722560.

Tillman, B., Fitts, D.J., Woodson, W.E., Rose-Sundholm, R. & Tillman, P. (2016). *Human Factors and Ergonomics Design Handbook* (3rd ed.). McGraw-Hill Education, New York. ISBN-13: 978-0071702874.

Tognazzini, B. (1992). *Tog on Interface*. Addison-Wesley Publishing Company, Inc., Reading, MA.

Tognazzini, B. (1993). Principles, Techniques, and Ethics of Stage Magic and Their Application to Human Interface Design. *Proceedings of Inter-CHI'93*, Amsterdam, The Netherlands, April 24–29. ACM, New York, pp. 355–362.

Travadel, S. & Guarnieri, F. (2015). L'agir en situation extrême (Acting in extreme situations). In *L'accident de Fukushima Daiichi: le récit du directeur de la centrale, volume 1, L'anéantissement*, F. Garnieri, Ed. Presses des Mines, Paris, France.

Tuegel, E.J., Ingraffea, A.R., Eason, T.G. & Spottswood, S.M. (2011). Reengineering aircraft structural life prediction using a digital twin. *International Journal of Aerospace Engineering*, 1(154798), pp. 1–14. doi:10.1155/2011/154798.

Vega-Mejia, C.A., Montoya-Torres, J.R. & Islam, S.M.N. (2016). Classification of Economic, Environmental and Social Factors in Vehicle Loading and Routing Operations. *ILS2016- Information Systems, Logistics and Supply Chain 6th International Conference*, Bordeaux, France.

von Bertalanffy, L. (1968). *General System Theory: Foundations, Development, Applications* (Revised ed.). Braziller, New York, NY, USA.

Warfield, J.N. (1971). *Societal Systems: Planning, Policy and Complexity*. Wiley, New York.

Watson, D.E. (2003). *Task Analysis: An Individual and Population Approach* (2nd ed.). American Occupational Therapy Association, Bethesda, MD.

Weber, M., Evans, J., Wolfson, M., DeLaura, R., Moser, B., Martin, B., Welch, J., Andrews, J. & Bertsimas, D. (2006). Improving air traffic management during thunderstorm. *12th Conference on Aviation, Range, and Aerospace Meteorology*, 29 January–2 February 2006.

Weibel, E.R. (1963). *Morphometry of the Human Lung*. Springer Verlag and Academic Press, New York. doi:10.1007/978-3-642-87553-3.

Wicks, R. (2011). Arthur Schopenhauer (Winter ed.), E.N. Zalta, Ed. The Stanford Encyclopedia of Philosophy, http://plato.stanford.edu/archives/win2011/entries/schopenhauer/.

Wiener, N. (1948). *Cybernetics: Or Control and Communication in the Animal and the Machine*. Hermann & Cie, MIT Press, Paris & Cambridge, Massachusetts. ISBN 978-0-262–73009-9; 1948, 2nd revised ed. 1961.

Wilson, G. (1992). Applied use of cardiac and respiratory measures. Practical considerations and precautions. *Biological Psychology*, 34(2–3), pp. 163–178.

Wilson, G. (2001). Real-time adaptive aiding using psychological operator state assessment. In *Engineering Psychology and Cognitive Ergonomics*, D. Harris, Ed. Ashgate, Aldershot, UK.

Wilson, R.A. & Keil, F.C. (Eds.) (2001). *The MIT Encyclopedia of the Cognitive Sciences*. MIT Press, Cambridge, MA, USA.

www.acad.bg/ebook/ml/Society%20of%20Mind.pdf

Yates, F.A. (1966). *The Art of Memory*. University of Chicago Press, Chicago. ISBN 102 269 50018.

Zola, E. (1886). *L'Œuvre*. Les Rougon-Macquart Series. Charpentier et Cie Editeurs, Paris.

Index

Printed in the United States
by Baker & Taylor Publisher Services